Teaching Mathematics:
Psychological Foundations

Teaching Mathematics: Psychological Foundations

F. Joe Crosswhite
Jon L. Higgins
Alan R. Osborne
Richard J. Shumway

The Ohio State University

Charles A. Jones Publishing Company
Worthington, Ohio

1 2 3 4 5 6 7 8 9 10 / 77 76 75 74 73

Library of Congress Catalog Card Number: 72-89108

International Standard Book Number: 0-8396-0029-1

Printed in the United States of America

Preface

Educators generally agree that knowledge of learning should provide a basis for decisions about teaching. Theories of learning have influenced curriculum and instruction in mathematics. The twentieth century history of emphases in mathematics education can be traced by a partial reflection of concurrent developments in psychology. However, decisions about teaching problems have varied according to the psychological allegiance of decision-makers. No single theory of learning has proved robust enough to encompass the range of ambiguities and complexities involed in mathematics instruction. The principles evolving from a single theory may work well in establishing computational skills and prove inadequate in developing concepts of proof. In the absence of a comprehensive theory to guide instruction, the teacher must borrow selectively from a variety of psychological theories those principles most relevant to the particular instructional decision to be made. *Mathematics Teaching: Psychological Foundations* provides a base for such an eclectic approach.

The articles selected center around the cognitive aspects of mathematics learning. They illustrate a variety of learning theories, emphasize differences between these theories, and relate the theories to mathematics instruction. The articles in Unit 1 present general analyses of learning in mathematics. They establish a framework within which the specific ideas in subsequent units can be placed in perspective. The nature of the learner is explored in Unit 2. The focus is on diagnosis as a prerequisite skill for successful prescription of remedial work. The intellectual growth and development of the child is examined as a complement to the diagnostic emphasis. Unit 3 is an extension and refinement of the major psychological approaches outlined in Unit 1. Each approach is tempered by the author's perception of the structure of mathematics. The specificity of the examples provides a set of principles directly applicable to decisions for mathematics instruction. While psychology helps a teacher select strategies for attaining the goals of instruction, determining an appropriate balance for the various types of objectives is a different task. Unit 4 related the decision-making processes involved in determining goals and in designing instructional strategies to achieve those goals.

Each reading is followed by a set of questions and activities which encourage the reader to apply the content of the article to broader situations or to contrast the content to other readings. Applying knowledge of principles of learning to instruction is exciting and demanding. Maintaining a comprehensive familiarity with theory is not enough; the effectiveness of each idea must be carefully tested and evaluated. Fitting principles of learning to the classroom situation demands continual creative problem-solving behavior by teachers. Each teacher makes assumptions about how he thinks his pupils learn. But most teachers make these assumptions at an implicit and intuitive level. One step toward improving teaching is to make these individual assumptions more explicit so they may be examined and evaluated. Knowledge of major psychological theories of learning and examination of the implications of these theories for mathematics teaching is one way teachers can become aware of their own assumptions about mathematics learning.

Mathematics Teaching: Psychological Foundations brings together a wide range of articles describing applications of psychological theories to mathematics learning and teaching. Preservice teachers can use these articles as a framework for observing and analyzing mathematics classrooms. Inservice teachers can try the classroom techniques suggested in these readings and evaluate their results with different individuals and classes. Students and researchers in mathematics education will find the readings a valuable source for formulating research questions and new theoretical constructs.

F. Joe Crosswhite

Jon L. Higgins

Alan R. Osborne

Richard J. Shumway

Contents

The Charles A. Jones Publishing Company

International Series in Education

Adams, *Simulation Games*

Allen, Barnes, Reece, Roberson, *Teacher Self-Appraisal: A Way of Looking Over Your Own Shoulder*

Armstrong, Cornell, Kraner, Roberson, *The Development and Evaluation of Behavioral Objectives*

Braun, Edwards, *History and Theory of Early Childhood Education*

Carlton, Goodwin, *The Collective Dilemma: Negotiations in Education*

Criscuolo, *Improving Classroom Reading Instruction*

Crosswhite, Higgins, Osborne, Shumway, *Mathematics Teaching: Psychological Foundations*

Denues, *Career Perspective: Your Choice of Work*

DeStefano, *Language, Society, and Education: A Profile of Black English*

Doll, *Leadership to Improve Schools*

Drier, *K-12 Guide for Integrating Career Development into Local Curriculum*

Foster, Fitzgerald, Beal, *Career Education and Vocational Guidance*

Frymier, Hawn, *Curriculum Improvement for Better Schools*

Goodlad, Klein, Associates, *Behind the Classroom Door*

Hauenstein, *Curriculum Planning for Behavioral Development*

Higgins, *Mathematics Teaching and Learning*

Hitt, *Education as a Human Enterprise*

Leland, Smith, *Mental Retardation: Perspectives for the Future*

Lutz, *Toward Improved Urban Education*

Meyer, *A Statistical Analysis of Behavior*

National Society for the Study of Education, *Contemporary Educational Issues* (10 book series)

Nerbovig, *Unit Planning: A Model for Curriculum Development*

Overly, Kinghorn, Preston, *The Middle School: Humanizing Education for Youth*

Perry, Wildman, *The Impact of Negotiations in Public Education: The Evidence from the Schools*

Poston, *Implementing Career Education*

Pula, Goff, *Technology in Education: Challenge and Change*

Ressler, *Career Education: The New Frontier*

Rich, *Humanistic Foundations of Education*

Shane, Shane, Gibson, Munger, *Guiding Human Development: The Counselor and the Teacher in the Elementary School*

Swanson, *Evaluation in Education*

Thiagarajan, *The Programing Process: A Practical Guide*

Von Haden, King, *Innovations in Education: Their Pros and Cons*

Weber, *Early Childhood Education: Perspectives on Change*

Wernick, *Career Education in the Elementary School*

Wiles, *Changing Perspectives in Educational Research*

Wiman, *Instructional Materials*

Unit 1

An Introduction to Theories of Mathematics Learning

Each article in this section would serve well as an introduction to the readings in this book. However, each possesses a unique characteristic which can and should be translated into operational terms if one is to apply psychology in the mathematics classroom.

Shulman's article embodies a comprehensive framework for analyses of human learning. His treatment of competing theories is developed in terms of the philosophical and psychological bases for the models of learning. The contrast is sharpened by discussion of issues, problems, and goals endemic to classroom instruction; namely, discovery, expository teaching, transfer of learning, the role of the mathematical structure, and the characteristics of the learner. The generality of the article provides a useful comparative basis for reading the other articles in the book; you should turn to it frequently as you read other articles of a more specific nature.

The second article introduces another element into the framework for decision making. Gagné categorizes the kinds of learning expected in a mathematics classroom. Concept learning is differentiated from principle learning and problem solving. As you read you should create examples of the various types of learning. Consider whether you would implement different instructional strategies to facilitate these types of learning.

The remaining article exemplifies the thoughtful, problem-solving attitude which must be a component of the teacher's behavior as he uses psychology to increase his control over the learning environment. Cronbach's article addresses a number of open questions in educational psychology. Many of these questions relate directly to emerging patterns in the mathematics curriculum. The sketch of historical developments not only highlights the shifting pattern of beliefs affecting mathematics teaching, but also questions the surety of belief in some well-established principles of instruction. His use of research to illustrate psychological properties is consistent with the scientific orientation teachers should exhibit. The discussion of the impact of the recent patterns of objectives provides a particularly appropriate frame of reference as you read further in this book.

Psychological Controversies in the Teaching of Science and Mathematics

Lee S. Shulman

The popular press has discovered the discovery method of teaching. It is by now, for example, an annual ritual for the Education section of *Time* magazine to sound a peal of praise for learning by discovery (e.g., *Time,* December 8, 1967). (7) *Time's* hosannas for discovery are by no means unique, reflecting as they do the educational establishment's general tendency to make good things seem better than they are. Since even the soundest of methods can be brought to premature mortality through an overdose of unremitting praise, it becomes periodically necessary even for advocates of discovery, such as I, to temper enthusiasm with considered judgment.

The learning by discovery controversy is a complex issue which can easily be oversimplified. A recent volume has dealt with many aspects of the issue in great detail. (8) The controversy seems to center essentially about the question of how much and what kind of guidance ought to be provided to students in the learning situation. Those favoring learning by discovery advocate the teaching of broad principles and problem solving through minimal teacher guidance and maximal opportunity for exploration and trial-and-error on the part of the student. Those preferring guided learning emphasize the importance of carefully sequencing instructional experiences through maximum guidance and stress the importance of basic associations or facts in the service of the eventual mastering of principles and problem solving.

Needless to say, there is considerable ambiguity over the use of the term *discovery.* One man's discovery approach can easily be confused with another's guided learning curriculum if the unwary observer is not alerted to the preferred labels ahead of time. For this reason I have decided to contrast the two positions by carefully examining the work of two men, each of whom is considered a leader of one of these general schools of thought.

Professor Jerome S. Bruner of Harvard University is undoubtedly the single person most closely identified with the learning-by-discovery position. His book, *The Process of Education,* captured the spirit of discovery in the new mathematics and science curricula and communicated it effectively to professionals and laymen.(1) His thinking

Lee S. Shulman, "Psychological Controversies in the Teaching of Science and Mathematics," *The Science Teacher* (September, 1968): 34-38.

Invited paper to the American Association for the Advancement of Science, Division Q (Education), New York City. December 1967.

will be examined as representative of the advocates of discovery learning.

Professor Robert M. Gagné of the University of California is a major force in the guided learning approach. His analysis of *The Conditions of Learning* is one of the finest contemporary statements of the principles of guided learning and instruction. (3)

I recognize the potential danger inherent in any explicit attempt to polarize the positions of two eminent scholars. My purpose is to clarify the dimensions of a complex problem, not to consign Bruner and Gagné to irrevocable extremes. Their published writings are employed merely to characterize two possible positions on the role of discovery in learning, which each has expressed eloquently at some time in the recent past.

In this paper I will first discuss the manner in which Bruner and Gagné, respectively, describe the teaching of some particular topic. Using these two examples as starting points, we will then compare their positions with respect to instructional objectives, instructional styles, readiness for learning, and transfer of training. We will then examine the implications of this controversy for the process of instruction in science and mathematics and the conduct of research relevant to that process.

Instructional Example: Discovery Learning

In a number of his papers, Jerome Bruner uses an instructional example from mathematics that derives from his collaboration with the mathematics educator, Z. P. Dienes. (2)

A class is composed of eight-year-old children who are there to learn some mathematics. In one of the instructional units, children are first introduced to three kinds of flat pieces of wood or "flats." The first one, they are told, is to be called either the "unknown square" or "X square." The second flat, which is rectangular, is called "1 X" or just X, since it is X long on one side and 1 long on the other. The third flat is a small square which is 1 by 1, and is called 1.

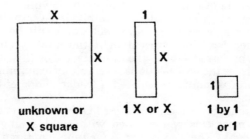

After allowing the children many opportunities simply to play with these materials and to get a feel for them, Bruner gives the children a problem. He asks them, "Can you make larger squares than this X square by using as many of these flats as you want?" This is not a difficult task for most children and they readily make another square.

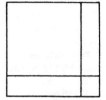

Bruner then asks them if they can describe what they have done. They might reply, "We have one square X, with two X's and a 1." He then asks them to keep a record of what they have done. He may even suggest a notational system to use. The symbol $X^\square$ could represent the square X, and a + for "and." Thus, the pieces used could be described as $X^\square + 2X + 1$.

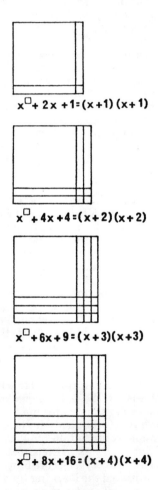

$$x^\square + 2x + 1 = (x+1)(x+1)$$

$$x^\square + 4x + 4 = (x+2)(x+2)$$

$$x^\square + 6x + 9 = (x+3)(x+3)$$

$$x^\square + 8x + 16 = (x+4)(x+4)$$

Another way to describe their new square, he points out, is simply to describe each side. With an X and a 1 on each side, the side can be described as X + 1 and the square as (X + 1) (X + 1) after some work with parentheses. Since these are two basic ways of describing the same square, they can be written in this way: $X^{\square} + 2X + 1 = (X + 1)(X + 1)$. This description, of course, far oversimplifies the procedures used.

The children continue making squares and generating the notation for them.

At some point Bruner hypothesizes that they will begin to discern a pattern. While the X's are progressing at the rate of 2, 4, 6, 8, the ones are going 1, 4, 9, 16, and on the right side of the equation the pattern is 1, 2, 3, 4. Provocative or leading questions are often used Socratically to elicit this discovery. Bruner maintains that, even if the children are initially unable to break the code, they will sense that there is a pattern and try to discover it. Bruner then illustrates how the pupils transfer what they have learned to working with a balance beam. The youngsters are ostensibly learning not only something about quadratic equations, but more important, something about the discovery of mathematical regularities.

The general learning process described by Bruner occurs in the following manner: First, the child finds regularities in his manipulation of the materials that correspond with intuitive regularities he has already come to understand. Notice that what the child does for Bruner is to find some sort of match between what he is doing in the outside world and some models or templates that he already has in his mind. For Bruner, it is rarely something *outside* the learner that is discovered. Instead the discovery involves an internal reorganization of previously known ideas in order to establish a better fit between those ideas and the regularities of an encounter to which the learner has had to accommodate.

This is precisely the philosophy of education we associate with Socrates. Remember the lovely dialogue of the *Meno* by Plato, in which the young slave boy is brought to an understanding of what is involved in doubling the area of a square. Socrates maintains throughout this dialogue that he is not teaching the boy anything new; he is simply helping the boy reorganize and bring to the fore what he has always known.

Bruner almost always begins with a focus on the production and manipulation of materials. He describes the child as moving through three levels of representation. The first level is the *enactive level,* where the child manipulates materials directly. He then progresses to the *ikonic level,* where he deals with mental images of objects but does not manipulate them directly. Finally he moves to the *symbolic level,* where he is strictly manipulating symbols and no longer mental images of objects. This sequence is an outgrowth of the developmental work of Jean Piaget. The synthesis of these concepts of manipulation of actual materials as part of a developmental model and the Socratic notion of learning as internal reorganization into a learning-by-discovery approach is the unique contribution of Jerome Bruner.

The Process of Education was written in 1959, after most mathematics innovations that use discovery as a core had already begun. It is an error to say that Bruner initiated the learning-by-discovery approach. It is far more accurate to say that, more than any one man, he managed to capture its spirit, provide it with a theoretical foundation, and disseminate it. Bruner is not the discoverer of discovery; he is its prophet.

Instructional Example: Guided Learning

Robert Gagné takes a very different approach to instruction. He begins with a task analysis of the instructional objectives. He always asks the question, "What is it you want the learner to be able to do?" This *capability* he insists, must be stated *specifically* and *behaviorally*.

By capability, he means the ability to perform certain specific functions under specified conditions. A capability could be the ability to solve a number series. It might be the ability to solve some problems in nonmetric geometry.

This capability can be conceived of as a terminal behavior and placed at the top of what will eventually be a complex pyramid. After analyzing the task, Gagné asks, "What would you need to know in order to do that?" Let us say that one could not complete the task unless he could first perform prerequisite tasks *a* and *b*. So a pyramid begins.

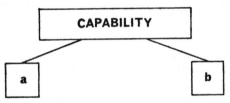

But in order to perform task *a*, one must be able to perform tasks *c* and *d* and for task *b*, one must know *e, f,* and *g.*

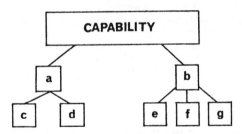

So one builds a very complex pyramid of prerequisites to prerequisites to the objective which is the desired capability.

Gagné has developed a model for discussing the different levels of such a hierarchy. If the final capability desired is a *problem-solving* capability, the learner first must know certain *principles*. But to understand those principles, he must know specific *concepts,* and prerequisite to these are particular *simple associations* or *facts* discriminated from each other in a distinctive manner. He continues the analysis until he ends up with the fundamental building blocks of learning—classically or operantly conditioned responses.

Gagné, upon completing the whole map of prerequisites, would administer pretests to determine which have already been mastered.

Upon completing the diagnostic testing, the resulting pattern identifies precisely what must be taught. This model is particularly conducive to subsequent programing of materials and programed instruction. When prerequisites are established, a very tight teaching program or package develops.

Earlier, we discussed the influences on Bruner. What influenced Gagné? This approach to teaching comes essentially from a combination of the neobehaviorist psychological tradition and the task analysis model that dominates the fields of military and industrial training. It was precisely this kind of task analysis that contributed to successful programs of pilot training in World War II. Gagné was trained in the neobehaviorist tradition and spent the major portion of his early career as an Air Force psychologist.

Nature of Objectives

The positions of Bruner and Gagné take very different points of view with respect to the objectives of education. This is one of the major reasons why most attempts at evaluating the relative effectiveness of these two approaches have come to naught. They really cannot agree on the same set of objectives. Any attempt to ask which is better—Michigan State's football team or the Chicago White Sox—will never succeed. The criteria for success are different, and it would be absurd to have them both on the same field competing against each other.

For Gagné, or the programed-instruction position which can be derived from him, the objectives of instruction are capabilities. They are behavioral products that can be specified in operational terms. Subsequently they can be task-analyzed; then they can be taught. Gagné would subscribe to the position that psychology has been successful in suggesting ways of teaching only when objectives have been made operationally clear. When objectives are not clearly stated, the psychologist can be of little assistance. He insists on objectives clearly stated in behavioral terms. They are the cornerstones of his position.

For Bruner, the emphasis is quite different. The emphasis is not on the *products* of learning but on the *processes*. One paragraph from *Toward a Theory of Instruction* captures the spirit of educational objectives for Bruner. After discussing the mathematics example previously mentioned, he concludes,

> Finally, a theory of instruction seeks to take account of the fact that a curriculum reflects not only the nature of knowledge itself—the specific capabilities—but also the nature of the knower and of the knowledge getting process. It is the enterprise par excellence where the line between the subject matter and the method grows necessarily indistinct. A body of knowledge, enshrined in a university faculty, and embodied in a series of authoritative volumes is the result of much prior intellectual activity. To instruct someone in these disciplines is not a matter of getting him to commit the results to mind; rather, it is to teach him to participate in the process that makes possible the establishment of knowledge. We teach a subject, not to produce little living libraries from that subject, but rather to get a student to think mathematically for himself, to consider matters as a historian does, *to take part in the process of knowledge-getting. Knowing is a process, not a product.* (2: 72) [Italics mine]

Speaking to the same issue, Gagné's position is clearly different.

> Obviously, strategies are important for problem solving, regardless of the content of the problem. The suggestion from some writings is that they are of overriding importance as a goal of education. After all, should not formal instruction in the school have the aim of teaching the student "how to think"? If strategies were deliberately taught, would not this produce people who could then bring to bear superior problem solving capabilities to any new situation. Although no one would disagree with the aims expressed, it is exceedingly doubtful that they can be brought about by teaching students "strategies" or "styles" of thinking. Even if these could be taught (and it is possible that they could), they would not provide the individual with the basic firmament of thought, which is subject-matter knowledge. Knowing a set of strategies is not all that is required for thinking; it is not even a substantial part of what is needed. *To be an effective problem solver, the individual must somehow have acquired masses of structurally organized knowledge. Such knowledge is made up of content principles, not heuristic ones.* (3: 170) [Italics mine]

While for Bruner "knowing is a process, not a product," for Gagné, "knowledge is made up of content principles, not heuristic ones." Thus, though both espouse the acquisition of knowledge as the major objective of education, their definitions of *knowledge* and *knowing* are so disparate that the educational objectives sought by each scarcely overlap. The philosophical and psychological sources of these differences will be discussed later in this paper. For the moment, let it be noted that when two conflicting approaches seek such contrasting objectives, the conduct of comparative educational studies becomes extremely difficult.[1]

Instructional Styles

Implicit in this contrast is a difference in what is meant by the very words *learning by discovery*. For Gagné, *learning* is the goal. How a behavior or capability is learned is a function of the task. It may be by discovery, by guided teaching, by practice, by drill, or by review. The focus is on *learning* and discovery is but one way to learn something. For Bruner, it is learning *by discovery*. The method of learning is the significant aspect.

For Gagné, in an instructional program the child is carefully guided. He may work with programed materials or a programed teacher (one who follows quite explicitly a step-by-step guide). The child may be quite active. He is not necessarily passive; he is doing things, he is working exercises, he is solving problems. But the sequence is determined entirely by the program. (Here the term "program" is used in a broad sense, not necessarily simply a series of frames.)

For Bruner much less system or order is necessary for the package, although such order is not precluded. In general, Bruner insists on the

---

[1]Gagné has modified his own position somewhat since 1965. He would now tend to agree, more or less, with Bruner on the importance of processes or strategies as objectives of education. He has not, however, changed his position regarding the role of sequence in instruction, the nature of readiness, or any of the remaining topics in this paper. (5) The point of view concerning specific behavioral products as objectives is still espoused by many educational theorists and Gagne's earlier arguments are thus still relevant as reflections of that position.

child manipulating materials and dealing with incongruities or contrasts. He will always try to build potential or emergent incongruities into the materials. Robert Davis calls this operation "torpedoing" when it is initiated by the teacher. He teaches a child something until he is certain the child knows it. Then he provides him with a whopper of a counterexample. This is what Bruner does constantly—providing contrasts and incongruities in order to get the child, because of his discomfort, to try to resolve this disequilibrium by making some discovery (cognitive restructuring). This discovery can take the form of a new synthesis or a new distinction. Piaget, too, maintains that cognitive development is a process of successive disequilibria and equilibria. The child, confronted by a new situation, gets out of balance and must accommodate to achieve a new balance by modifying the previous cognitive structure.

Thus, for Gagné, instruction is a smoothly guided tour up a carefully constructed hierarchy of objectives; for Bruner, instruction is a roller-coaster ride of successive disequilibria and equilibria until the desired cognitive state is reached or discovered.

Readiness

The guided learning point of view, represented by Gagné, maintains that readiness is essentially a function of the presence or absence of prerequisite learning.

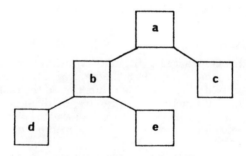

When the child is capable of *d* and *e* above, he is by definition ready to learn *b*. Until then he is not ready. Gagné is not concerned with genetically developmental considerations. If the child at age five does not have the concept of the conservation of liquid volume, it is not because of an unfolding in his mind; he just has not had the necessary prior experiences. Ensure that he has acquired the prerequisite behaviors, and he will be able to conserve. (4)

For Piaget (and Bruner) the child is a developing organism, passing through cognitive stages that are biologically determined. These stages are more or less age-related, although in different cultures certain stages may come earlier than others. To identify whether the child is ready to learn a particular concept or principle, one analyzes the structure of that to be taught and compares it with what is already known about the cognitive structure of the child at that age. If they are consonant, it can be taught; if they are dissonant, it cannot.

Given this characterization of the two positions on readiness, to which one would you attribute the following statement? ". . . any subject can be taught effectively in some intellectually honest form to any child at any stage of development." While it sounds like Gagné, you recognize that it isn't—it's Bruner! (2: 33) And in this same chapter he includes an extensive discussion of Piaget's position. Essentially he is attempting to translate Piaget's theories into a psychology of instruction.

Many are puzzled by this stand, including Piaget. In a recent paper delivered in the United States, he admitted that he did not understand how Bruner could make such a statement in the light of Piaget's experiments. If Bruner meant the statement literally; i.e., *any* child can learn *anything*, then it just is not true! There are always things a child cannot learn, especially not in an intellectually honest way. If he means it homiletically, i.e., we can take almost anything and somehow resay it, reconstruct it, restructure it so it now has a parallel at the child's level of cognitive functioning, then it may be a truism.

I believe that what Bruner is saying, and it is neither trivial nor absurd, is that our older conceptions of readiness have tended to apply Piagetian theory in the same way as some have for generations applied Rousseau's. The old thesis was, "There is the child—he is a developing organism, with invariant order, invariant schedule. Here, too, is the subject matter, equally hallowed by time and unchanging. We take the subject matter as our starting point, watch the child develop, and feed it in at appropriate times as he reaches readiness." Let's face it; that has been our general conception of readiness. We gave reading readiness tests and hesitated to teach the pupil reading until he was "ready." The notion is quite new that the reading readiness tests tell not when to begin teaching the child, but rather what has to be done to get him more ready. We used to just wait until he got ready. What Bruner is suggesting is that we must modify our conception of readiness so that it includes not only the child but the subject matter. Subject matter, too, goes through stages of readiness. The same subject matter can be represented at a manipulative or enactive level, at an ikonic level, and finally at a symbolic or formal level. The resulting model is Bruner's concept of a spiral curriculum.

Piaget himself seems quite dubious over the attempts to accelerate cognitive development that are reflected in many modern math and science curricula. On a recent trip to the United States, Piaget commented.

> . . . we know that it takes nine to twelve months before babies develop the notion that an object is still there even when a screen is placed in front of it. Now kittens go through the same stages as children, all the same substages, but they do it in three months—so they're six months ahead of babies. Is this an advantage or isn't it? We can certainly see our answer in one sense. The kitten is not going to go much further. The child has taken longer, but he is capable of going further, so it seems to me that the nine months probably were not for nothing.
>
> It's probably possible to accelerate, but maximal acceleration is not desirable. There seems to be an optimal time. What this optimal time is will surely depend upon each individual and on the subject matter. We still need a great deal of research to know what the optimal time would be. (6: 82)

The question that has not been answered, and which Piaget whimsically calls the "American question," is the empirical experimental question: To what extent it is possible through a Gagnéan approach to accelerate what Piaget maintains is the invariant clockwork of the order? Studies being conducted in Scandinavia by Smedslund and in this country by Irving Sigel, Egon Mermelstein, and others are attempting to identify the degree to which such processes as the principle of conservation of volume can be accelerated. If I had to make a broad generalization, I would have to conclude that at this point, in general, the score for those who say you cannot accelerate is somewhat higher than the score for those who say that you can. But the question is far from resolved; we need many more inventive attempts to accelerate cognitive development than we have had thus far. There remains the question of whether such attempts at experimental acceleration are strictly of interest for psychological theory, or have important pedagogical implications as well—a question we do not have space to examine here.

Sequence of the Curriculum

The implications for the sequence of the curriculum growing from these two positions are quite different. For Gagné, the highest level of learning is problem solving; lower levels involve facts, concepts, principles, etc. Clearly, for Gagné, the appropriate sequence in learning is, in terms of the diagram below, from the bottom up. One begins with simple prerequisites and works up, pyramid fashion, to the complex capability sought.

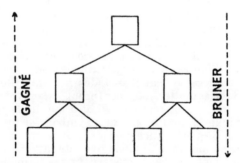

For Bruner, the same diagram may be appropriate, but the direction of the arrow would be changed. He has a pupil begin with *problem solving*. This process is analogous to teaching someone to swim by throwing him into deep water. The theory is that he will learn the fundamentals because he needs them. The analogy is not totally misbegotten. In some of the extreme discovery approaches we lose a lot of pupils by mathematical or scientific drowning. As one goes to the extreme of this position, he runs the risk of some drownings. For Gagné, the sequence is from the simple to the complex; for Bruner one starts with the complex and plans to learn the simple components in the context of working with the complex.

It is unclear whether Bruner subscribes to his position because of his concept of the nature of learning or for strictly motivational reasons. Children may be motivated more quickly when given a problem they cannot solve, than they are when given some little things to learn on the promise that if they learn these well, three weeks from now they will be able to solve an exciting problem. Yet, Bruner clearly maintains that learning things in this fashion also improves the transferability of what is learned. It is to a consideration of the issue of transfer of training that we now turn.

Transfer of Training

To examine the psychologies of learning of these two positions in any kind of comprehensive form would require greater attention than can be devoted here, but we shall consider one concept—that of transfer of training. This is probably the central concept, or should be, in any educationally relevant psychology of learning.

Gagné considers himself a conservative on matters of transfer. He states that "transfer occurs because of the occurrence of specific identical (or highly similar) elements within developmental sequences." (4: 20) To the extent that an element which has been learned, be it association, concept, or principle, can be directly employed in a new situation, transfer will occur. If the new context requires a behavior substantially different from the specific capability mastered earlier, there will be no transfer.

Bruner, on the other hand, subscribes to the broadest theories of transfer of training. Bruner believes that we can have massive transfer from one learning situation to another. Broad transfer of training occurs when one can identify in the structures of subject matters basic, fundamentally simple concepts or principles which, if learned well, can be transferred both to other subject matters within that discipline and to other disciplines as well. He gives examples such as the concept of conservation or balance. Is it not possible to teach balance of trade in economics in such a way that when ecological balance is considered, pupils see the parallel? This could then be extended to balance of power in political science, or to balancing equations.

Even more important, for Bruner, is the broad transferability of the knowledge-getting processes—strategies, heuristics, and the like—a transfer whose viability leaves Gagné with deep feelings of doubt. This is the question of whether learning by discovery leads to the ability *to* discover, that is, the development of broad inquiry competencies in students.

What does the evidence from empirical studies of this issue seem to demonstrate? The findings are not all that consistent. I would generalize them by saying that most often guided learning or expository sequences seem to be superior methods for achieving immediate learning. With regard to long-term retention, the results seem equivocal, with neither approach consistently better. Discovery learning approaches appear to be superior when the criterion of transfer of principles to new situations is employed. (9) Notably absent are studies which deal with the question of whether general techniques, strategies, and heuristics of discovery can be learned—by discovery or in any other manner—which will transfer across grossly different kinds of tasks.

Why is transfer of training superior in the discovery situation when the learning of principles is involved? There are two kinds of transfer—positive transfer and negative transfer. We call something positive transfer when mastery of task X facilitates mastery of task Y. Negative transfer occurs when mastery of task X inhibits mastery of task

Y. Positive transfer is a familiar notion for us. Negative transfer can be exemplified by a piece of advice baseball coaches often give their players. They tell them not to play golf during the baseball season because the baseball swing and the golf swing involve totally different muscles and body movements. Becoming a better golf swinger interferes with the baseball swing. In psychological terms there is negative transfer between golf and baseball.

What is needed for positive transfer is to minimize all possible interference. In transfer of training, there are some ways in which the tasks transferred to are like the ones learned first, but in other ways they are different. So transfer always involves striking a balance between these conflicting potentials for both positive and negative transfer. In discovery methods, learners may transfer more easily because they learn *the immediate things less well.* They may thus learn the broad strokes of a principle, which is the aspect most critical for remote transfer, while not learning well the detailed application of that specific principle, which could interfere somewhat with successful remote transfer.

If this formulation is correct, we are never going to find a method that will both allow for tremendous specific learning of products and broad transfer, because we are dealing in a closed system in which one must make a choice. To the extent that initial learning is well done, transfer is restricted. The instructor may have to decide which is more important—an immediate specific product or broad transfer—and choose his subsequent teaching method on the basis of that decision. This is a pessimistic view, and I hope that future studies might find it flawed.

Synthesis or Selection

Need we eternally code these as two alternatives—discovery versus expository teaching—or can we, without being heretical, manage to keep both of these in our methodological repertories as mathematics and science educators?

John Dewey was always very suspicious whenever he approached a controversy between two strongly stated positions, each of which insisted that the other was totally in error. The classic example of this is in his monograph *Experience and Education,* in which he examines the controversy of traditional versus progressive education. Dewey teaches us that whenever we confront this kind of controversy, we must look for the possibility that each position is massively buttressed by a brilliant half-truth from which is extrapolated the whole cloth of an educational philosophy. That is, too often a good idea wears thin as its advocates insist that it be applied outside its appropriate domain.

As educators, we find it extremely important to identify the conditions under which each of these theories can be applied most fruitfully. First, one must examine the nature of the objectives. More than half of this controversy can be resolved not at the level of which is the better psychology, but at the level of evaluative philosophical judgments. Given one set of goals, clearly the position Gagné advocates presently has more evidence in its favor; given another set of goals, there is no question but that Bruner's position is preferable to Gagné's.

But there are other questions. The age and personality of the learner must be taken into account. All things being equal, there are some kinds of children who cannot tolerate the ambiguity of a discovery experience.

We all know this; some of us prefer to hear lectures that are well organized so that we can take notes in a systematic manner. Others of us like nothing better than a free-flowing bull session; and each of us is convinced that we learn more in our preferred mode than the others learn in theirs. Individual differences in learning styles are major determinants of the kinds of approaches that work best with different children.

Yet this is something we have in general not taken into consideration at all in planning curricula—and for very good reasons. As yet, we do not have any really valid ways of measuring these styles. Once we do, we will have a powerful diagnostic tool. Subject matter, objectives, characteristics of children, and characteristics of the teacher are all involved in this educational decision. Some teachers are no more likely to conduct a discovery learning sequence than they are to go frugging at a local nightclub.

There appear to be middle routes as well. In many of the experimental studies of discovery learning, an experimental treatment labeled *guided discovery* is used. In guided discovery, the subjects are carefully directed down a particular path along which they are called upon to discover regularities and solutions on their own. They are provided with cues in a carefully programed manner, but the actual statement of the principle or problem solution is left up to them. Many of the well-planned Socratic dialogues of our fine teachers are forms of guided discovery. The teacher carefully leads the pupils into a series of traps from which they must now rescue themselves.

In the published studies, guided discovery treatments generally have done quite well both at the level of immediate learning and later transfer. Perhaps this approach allows us to put the Bruner roller-coaster of discovery on the well-laid track of a Gagné hierarchy.

Thus, the earlier question of which is better, learning by discovery or guided learning, now can be restated in more functional and pragmatic terms. Under what conditions are each of these instructional approaches, some sequence or combination of the two, of some synthesis of them, most likely to be appropriate? The answers to such questions ought to grow out of quite comprehensive principles of human learning. Where are we to find such principles?

Theories of Learning and the Science and Mathematics Curriculum

There is a growing psychology of learning that is finally becoming meaningful to curriculum construction and educational practice. Children are being studied as often as rats, and classrooms as often as mazes. Research with lower animals has been extremely useful in identifying some principles of learning that are so basic, so fundamental, so universal that they apply to any fairly well-organized blob of protoplasm. But there is a diminishing return in this approach insofar as transfer to educational practice is concerned. Today, a developing, empirically based psychology of learning for *homo sapiens* offers tremendous promise. But it can never be immediately translatable into a psychology of the teaching of mathematics or science. Mathematics and science educators must not make the mistake that the reading people have made and continue to make. The reason that the psychology of the teaching of reading has made such meager progress in the last twenty-five years is that the reading people have insisted on being borrowers. Something new happens in linguistics and within three years a linguistic reading series is off the press. It is an attempt to bootleg an idea from one field and put it directly into another without the necessary intervening steps of empirical testing and research.

Mathematics and science education are in grave danger of making that same error, especially with the work of Piaget and Bruner. What is needed now are well-developed empirically based psychologies of mathematics and science learning. Surely they will grow out of what is already known about the psychology of learning in general, but they must necessarily depend upon people like yourselves, your students, and your colleagues who are interested in mathematics and science conducting empirical studies of how certain specific concepts are learned under certain specific conditions with certain specific kinds of pupils. If anything is true about the field of mathematics and science education today, it is that rarely have any disciplines been so rich in theory and brilliant ideas. But we must seriously consider the admonition of Ivan Pavlov, the great Russian psychologist, who is said to have told his students the following:

> Ideas and theories are like the wings of birds; they allow man to soar and to climb to the heavens. But facts are like the atmosphere against which those wings must beat, and without which the soaring bird will surely plummet back to earth.

References

1. Bruner, Jerome S. *The Process of Education.* Cambridge, Mass.: Harvard University Press, 1960.

2. Bruner, Jerome S. *Toward a Theory of Instruction.* Cambridge, Mass.: Belknap Press, 1966.

3. Gagné, Robert M. *The Conditions of Learning.* New York: Holt, Rinehart & Winston, 1965.

4. Gagné, Robert M. Contributions of Learning to Human Development. Address of the Vice-President, Section I (Psychology), American Association for the Advancement of Science, Washington, D. C. December 1966.

5. Gagné, Robert M. Personal communication. May 1968.

6. Jennings, Frank G. "Jean Piaget: Notes on Learning." *Saturday Review* (May 20, 1967): 82.

7. "Pain & Progress in Discovery." *Time* (December 8, 1967): 110 ff.

8. Shulman, Lee S. and Keislar, Evan R., eds. *Learning by Discovery: A Critical Appraisal.* Chicago: Rand-McNally, 1966.

9. Worthen, Blaine R. "Discovery and Expository Task Presentation in Elementary Mathematics." *Journal of Educational Psychology Monograph Supplement* 59: 1, Part 2; February 1968.

**Exercises: Psychological Controversies
in the Teaching of Science and Mathematics**

1. Is there a difference between teaching by discovery and learning by discovery? Can you distinguish between the teaching act and the learning act?

2. Is the discovery process different for a twelfth grader than it is for a first grader? If so, how? If not, why not?

3. Choose a single, simple concept from the mathematics curriculum. Describe a teaching strategy à la Bruner. Describe a teaching strategy à la Gagné. Identify the differences. The similarities.

4. Compare Bruner's and Gagné's positions with respect to the goals of instruction. Would they assign priorities of learning in the same way? Do they have the same goals? Explain. the same way? Do they have the same goals? Explain.

5. What is readiness? How do Bruner and Gagné differ with regard to readiness? Whose notion do you favor? Explain.

6. Bruner believes that *method* is the most important aspect of the learning process. Identify five topics in mathematics in which the method or process is a more appropriate learning goal than the specific outcome involved.

7. Gagné emphasizes that *learning* is the goal. Identify five topics in mathematics that depend more upon the knowledge of facts than upon a process. (Hint: consider the distinction between theorems and lemmas.)

8. Can a teacher who is following a sequence specified by a learning hierarchy teach by the discovery method at the same time? Why or why not?

9. Does a Gagné hierarchy depend entirely upon the structure of the content to be learned? Could it be different for different learners?

10. Compare the positions of Gagné and Bruner on transfer. How do they differ? Which position do you feel is closest to the belief of the average layman? Why? What is your position? How will you behave as a teacher because of your position?

11. Shulman indicates that one must minimize interference in order to achieve transfer. Name ways in which an elementary school child could find interference between the multiplication process and the procedure used for division. How might a teacher minimize this interference? Can you

find similar examples of topics that might produce interference in algebra? in geometry?

12. Give an example of a classroom mathematics topic. Illustrate in detail a Gagné approach and a Bruner approach. Describe the differences, if any, in objectives, teaching techniques, and evaluation.

Varieties
of Learning

Robert M. Gagné

. . . In discussing the various kinds of learning, it is necessary to refer to the components of the situation that are used to establish its presence, and occasionally also to employ simple diagrams to represent these components. These are a stimulus situation, represented by S, a set of responses, R, and the inferred connection between them, often shown by a set of connecting lines. These are simply convenient descriptive terms, and should not be considered to imply any of the varieties of "S-R theory," or other inferences concerning the mechanisms of behavior. The present consideration of the conditions of learning makes no attempt to evaluate the usefulness of these theories; it does not attempt to deal with them in any systematic fashion.

An observer of learning must deal with an input, an output, and a functioning entity in between. The input is a *stimulus situation* (S), which includes the varieties of changes in physical energy that reach the learner through his senses. Sometimes these stimuli are described in a shorthand way, for convenience, as when one describes a stimulus situation as "a projected slide image of the letter E," rather than specifying the brightness, size, and shape of this stimulus in physical terms. The stimulus situation is in general (except for the special instance of stimulation from the muscle sense) *outside the learner,* and can be identified and described in the terms of physical science.

The output, R, is also in a real sense outside the learner. It is a response or a set of responses that produces an identifiable product. The R that is the focus of interest here is not muscular movements, although it is well known that these are necessary for it. The output may be something like a "wave of the hand," but strictly speaking, it is not the muscular movements that underlie this event. Rather, it is the external, observable effect of these movements. The learner may produce a spoken output, which may leave no permanent record unless the observer of learning has used a recording machine; speaking the words is what is meant by R. Alternatively, the learner may write some words, and thus incidentally produce a permanent record. But the R is "writing words," as this may be seen by an observer. However, the record of these words may frequently be employed in judging the adequacy of this output.

The nature of the connection between an S and an R cannot be directly observed. Studies of nervous system functioning may some day provide a much greater understanding of these mechanisms than is now available. But in the meantime, it is desirable and necessary to make some immediate inferences (cf. Northrop, 1947) about the kind of transformation that must occur. To speculate about *how* such mechanisms work would be to construct a theory, as Pavlov, Hull, and other investigators have done, and which is not intended here. But simply to describe what the requirements must be in order for an observed transformation between S and R to occur is not constructing a theory, in the strict sense of the word. It may, however, be thought of as a description of what theories of learning have to explain, now or in the future. The kind of immediate inference that is based on observations of learning consists, then, of a statement (or diagram) indicating *what capability of the learner exists following the events of learning that was not there before.*

A simple example may be useful. A *reflex* is the inferred capability one is accustomed to make in describing the transformation between an S of a light shining on the retina, and an R of pupil contraction. In a diagram, such a simple transformation may be represented as a line, as

$$S \underline{\hspace{2cm}} R$$
light   pupil contraction

Obviously, the reflex is neither the S of the light nor the R of the contracting pupil; instead, it is the immediate inference we make about the nature of the transformation that occurs, the capability for which resides within the individual. This particular capability is of the sort that some authors have described as a one-to-one correlation; that is, whenever light (of specified brightness) shines on the retina, pupil contraction occurs. Of course, the reflex capability used here as an example is an unlearned one. But there are transformations of similar simplicity that are learned, and can therefore be represented in the same way. Also, as will be seen, there are more complex transformations that occur as a result of learning. . . .

Eight Types of Learning

Learning Type 1: Signal Learning

Learning to respond to a signal is a kind of learning quite familiar to everyone. Guthrie (1935, p. 48) gives this example of it:

> Two small country boys who lived before the day of the rural use of motor cars had their Friday afternoons made dreary by the regular visit of their pastor, whose horse they were supposed to unharness, groom, feed and water and then harness again on the departure. Their gloom was lightened finally by a course of action which one of them conceived. They took to spending the afternoon of the visit re-training the horse. One of them stood behind the horse with a hay-fork and periodically shouted "Whoa" and followed this with a sharp jab with the fork. Unfortunately no exact records of this experiment were preserved save that the boys were quite satisfied with the results.

This is, of course, a description of a set of conditions appropriate for the establishment of a *conditioned response*. It is customary to represent what has been learned here in the following way:

$$S \underline{\hspace{2cm}} R$$
"Whoa!"   pain response

In other words, a capability has been acquired by the horse that was not previously present: making a response typical of that to pain (including struggling, shying, running) to the sound of "Whoa!" It is as if the horse had learned to anticipate the painful stimulus; the verbal command has become a *signal* for pain. Some authors, notably Mowrer (1960a), propose that the response signaled in such situations as this is one of fear.

The conditions for establishment of this form of signal learning seem fairly clear. There must be nearly simultaneous presentation of two forms of stimulation: 1) the stimulus producing a generalized reaction of the sort one is interested in establishing, and 2) the stimulus providing the signal. If the "whoa" comes after the painful stimulus by a fraction of a second, it cannot become a signal; if it comes too many seconds before the painful stimulus, it also fails to produce the desired learning. The number of times this pairing of stimuli must occur is a question that has no single answer. There are many instances of signal learning of this type that have occurred in one trial, and there are many others that seem to require several pairings of the proper stimuli.

Obviously, such learning represents a somewhat specialized type, although it is not exactly rare in the occurrences of everyday existence. People who have peeled onions may feel tears at the sight of one. The young child may learn that a shout by one of his parents signals pain-to-come. The military command "Attention!" is designed to signal a condition of alertness. Involuntary fears of many sorts, such as fear of the water and fear of heights, may be engendered in individuals as children when these signals have been accompanied by painful or frightening stimulation. Presumably, pleasant emotions may also be involved in signal learning. The sight of the mother's face may become a learned signal to the infant for various pleasurable events associated with the presence of his mother. The sight of one's old high school after many years of absence may evoke pleasant feelings of nostalgia that are quite independent of the recall of any specific events.

All these examples, however, merely serve to emphasize one important characteristic of this type of signal learning (that is, the Pavlovian conditioned response): *the responses are general, diffuse, emotional ones.* This is learning that has a truly "involuntary" character, and applies to responses that are not typically under voluntary control. (cf. Kimble, 1961, pp. 44-108) A fear response, involving general, diffuse activity including speeded heartbeat, constriction of blood vessels, and other internal involuntary behavior may readily acquire a connection with a signal under the conditions we have described. But precise voluntary responses, such as kicking a football or writing one's name, cannot be acquired in this way.

Learning Type 2: Stimulus-Response Learning

Another type of learning to respond to a signal is probably much more important than the classical type we have just described. This is a kind of learning that involves making very precise movements of the skeletal muscles in response to very specific stimuli or combinations of stimuli. In the human being, we often speak of "voluntary responses" as constituting the observed output. In other words, this kind of learning makes it possible for an individual to perform an action when he wants to; so far as we know, the same is probably true of animals also.

A simple and well-known example of such learning occurs when a dog learns to "shake hands" in response to a vocal stimulus supplied by his master, or by another friendly person. The events that take place in such learning, as Mowrer (1960a, p. 215) describes them, are somewhat as follows: While playing with the puppy, the master says, "Shake hands!" At the same time, he gently raises the dog's paw and shakes it, then pats the dog's head, or gives him a piece of dog biscuit. He then repeats this entire procedure on several subsequent occasions, perhaps using progressively lighter force to raise the paw. After several occasions of this sort, the dog raises his own paw when his master says, "Shake hands," and is patted or fed as usual. Eventually, the dog comes to perform this act promptly and more or less precisely whenever the proper signal is given. It can then be said that the dog has learned what may be called a *stimulus-response* capability.

Obviously, this kind of learning is distinguishable from *signal learning* in terms of its outcome. The response acquired by this means is a fairly precise, circumscribed, skeletal muscular act, far different from the generalized emotional responding that characterizes the Pavlovian kind of signal-responding. In fact, as Kimble points out (1961), attempts to establish the latter type of learning by means of a sequence of events similar to that described for stimulus-response learning have generally been quite unsuccessful.

Returning to stimulus-response learning itself, it is important to take note of certain other conditions that attend the acquisition of this kind of capability. First, there appears to be a typical gradualness to the learning of this act. The dog does not suddenly shake hands, all at once; at least a few occasions of repetition appear to be necessary. Second, the response that the dog makes becomes more and more sure and precise throughout these several occasions. (The response is said by Ferster and Skinner (1957) to undergo "shaping.") Third, the controlling stimulus also becomes more and more precise—whereas initially the dog may respond to other commands than "Shake hands!" these other vocalizations eventually lose their control over the outcome. And finally, a point that many learning theorists would consider of utmost importance, there is *reward* or *reinforcement*. The dog is rewarded for responses that are "correct" or that approach being so in his master's view, and he is not rewarded for those that are incorrect.

The capability that has been acquired in this kind of situation ought to suggest this *differential* characteristic of the learning, by implying that a particular S-R is established, and at the same time other S-R's, equally probable at the beginning of learning, are disestablished. We can show this by using an arrow rather than the line between the S and the R, as $S \rightarrow R$, to emphasize that a *process of discrimination* is an integral part of this kind of learning. A degree of precision has been established in the response, which can easily be distinguished from similar although "wrong" responses. In like manner, there is precision in the stimulus, which is a particular complex of stimulation differing from other stimuli that may be present at the time the response is made.

Still another characteristic of stimulus-response learning must be noted before the description is complete. Every uncomplicated example of $S \rightarrow R$ learning indicates that it is *motor learning*. The implication of this statement is not simply to the effect that muscular movements are involved in the outcome, because this may be true of other varieties, including signal learning (as when an animal learns to struggle or run at a signal originally paired with shock). In $S \rightarrow R$ learning, though, an important component of the *stimulus itself* is generated by muscular movements. While the act is being established, the external stimulus "Shake hands!" is accompanied by proprioceptive stimulation from the muscles that raise the dog's paw. Even when the act is fully learned, some parts of this stimulation are still present. For example, the dog often raises his paw "as if voluntarily," even when no one has said "Shake hands." He may now "invite" his master to shake hands! Presumably, this portion of the total stimulation plays an important role in the learning process, as Mowrer (1960a) has emphasized. As will be seen, it is a very essential part of the capability that is learned.

This kind of learning can accordingly be represented in complete fashion as follows:

$$Ss \rightarrow R$$

where the S refers to an external signal, s the accompanying internal proprioceptive stimulation, and R the response. The arrow, rather than a line, is used to indicate the precise, discriminated nature of the capability; other potential $Ss \rightarrow R$'s are disestablished (or *extinguished*) by the events of learning.

In the descriptive language used by Skinner (1938), this capability is called a *discriminated operant*. A common example from the animal-learning laboratory is the rat learning to press a lever in the presence of a light (the external S), and at the same time extinghishing this response when the external S is darkness. The conditions for establishment are essentially the same as those we have described: 1) the lever is initially baited with food so that the rat will press it, and thus the presence of a proprioceptive s is ensured; 2) correct lever-pressing responses are reinforced by delivery of bits of food, while incorrect responses including those made in the absence of the light go unrewarded; 3) several repetitions are required to establish the capability.

This form of learning appears to govern the acquisition of a new vocalization habit by a young child and *can* be employed to teach an adult the pronunciation of an unfamiliar foreign word. The young child, being petted by the admiring parent, receives reinforcement for many of the responses he makes, including such things as smiling, gurgling, cooing, and vocalizing. The process of discrimination has already begun because some kinds of responses (such as crossing his eyes) are not likely to be rewarded. The parent may frequently introduce the desired signal, such as, "Say mama," and eventually the child does say, "Mama," more or less by chance, immediately following this signal. Suitable reinforcement follows, and the learning of this new capability is then well launched. It is noteworthy that nothing exactly comparable to lifting the dog's paw for him occurs here; the parent cannot directly manipulate the child's vocal cords. But the parent comes as close to this as he can by cleverly choosing the occasions for the stimulus "Say mama" to correspond with those times when the child is making (or about to make) movements of opening and shutting his mouth while breathing out and vocalizing. The learning conditions for this particular example are therefore somewhat inefficiently arranged, and the parent must make up for this by his astuteness in *selecting* the occasions when an appropriate internal s will occur.

Of course, teaching an adult to pronounce a new foreign word is a somewhat easier process, although much the same set of events is involved. The adult, though, begins with an *approximate* capability already established. When the signal is given, "Say *femme*," he can immediately say something that is almost correct. Subsequent trials in his case, therefore, are largely a matter of bringing about discrimination. He must receive reinforcement for responding to a narrow range of correct external sounds of *femme,* and also for a narrow range of internal stimuli from his muscles in pronouncing the word. The adult learner also reinforces himself, by recognizing a match between his vocalization of *femme* and that of his French teacher. If he is going to do this effectively, he must, of course, have previously adopted a suitably precise set of matching criteria.

Learning Type 3: Chaining

Another extremely simple and widely occurring learning situation is called *chaining.* In brief terms, it is a matter of connecting together in a sequence two (or more) previously learned $Ss \rightarrow R$'s. Our language is filled with such chains of verbal sequences, as is well known to students of linguistics, and is also revealed by studies of word association. "Horse and buggy," "boy meets girl," "never again" are three of the hundreds of examples that could be given in which the first member of the sequence seems firmly tied to the second.

It is relatively easy to arrange a set of conditions in which a dog who has learned to shake hands offers his paw after he barks, or alternatively, offers his paw and then barks. But perhaps one of the simplest human examples is that of a child learning to ask for a specific object by name.

Many a fond parent has attempted to teach his infant to call for an object, say a doll, by its name. After a number of parental tries at presenting the doll and saying "doll," distributed perhaps over several months, the child eventually achieves success. In fact, he appears to acquire such a capability "suddenly," and without there being an entirely clear relationship between his calling for a doll and the events that have gone before. Of course, growth factors are at work during this period: one cannot expect the nervous system of a newborn child to be ready to learn such a chain. But there are other aspects of readiness that are even more relevant to our present discussion, and probably also more important for an understanding of this kind of learned behavior.

Obviously, if the child is going to learn to ask for a doll, he must first know how to make this verbal response. This is a matter of stimulus-response learning (type 2), as we have seen. The sight and feel of the doll accompanied by the spoken word "doll" becomes the stimulus situation that is connected by learning with the child's saying "Doll." Thus there is established the stimulus-response sequence:

$$Ss \longrightarrow R$$
$$\text{doll} \qquad \text{"doll"}$$

Another $Ss \rightarrow R$ connection has also been learned, perhaps even earlier. The child has picked up the doll, handled it, hugged it, shaken it, and so on. Thus in a fundamental (and nonverbal) sense, he "knows what a doll is." A doll is the particular set of stimuli connected with hugging, and is thus distinguished from other stimuli, like a ball or a wagon, that cannot be hugged. Perhaps the hugging of the doll has been associated particularly with lying in his bed, preparatory to going to sleep. There is, then, some such connection as the following:

$$Ss \longrightarrow R$$
lying down   hugging doll

If both $Ss \rightarrow R$ connections are present, the chain of asking for the doll is a relatively simple matter to establish. First, the child is placed in his crib. Perhaps he makes the movements of hugging the doll that is not there. But now his mother shows him the doll; perhaps she also says, "Do you want your doll?" These events establish a sequence that sets the occasion for learning that may be depicted as follows:

$$Ss \longrightarrow R \quad \sim \quad Ss \longrightarrow R$$
lying down   hugging doll   doll   "doll"

The capability that has been learned is simply this: The child, on lying in his crib and seeing the doll outside the crib, "asks for" it by saying "Doll." He may even come to do so without actually seeing the doll, since the hugging responses he makes may generate internal stimuli (small s's that are a part of the stimulus situation of feeling the doll) that have themselves been previously connected with the vocalization "doll."

In a similar manner, the chain can be extended. Perhaps the child is initially placed in his crib in an upright position. His mother now says, "Lie down," and he has previously learned to respond to this stimulus in an S-R manner. Having lain down, the next connection in the sequence comes into play: he makes the incipient hugging movements toward his absent doll, and this in turn sets off the next link in the chain—he asks for his doll.

Now all these events are likely to occur in a very natural manner in the natural world. They occur so naturally, in fact, that it may not be apparent that any learning has occurred at all. Yet, at one period in time these chains are not there, and in a subsequent period, as if by magic, they *are* there! The magic, though, may be made to happen deliberately if we want it to, as may be true in teaching a child a chain like tying a knot or printing a letter. These are the conditions that appear to be necessary:

1. The individual links in the chain must have been previously established. The child must already have learned to say "Doll," to hug the doll, to lie down when his mother tells him to. (Of course, one *can* attempt to establish all these links at a single time, but that is not an efficient procedure.)

2. There must be *contiguity* of each link with the next following one. The hugging responses must be followed in a brief period by the stimulus leading to the child's vocalization "Doll," or the chain will not be established. Often such contiguity is "built into" the chain of events, as is the case, for example, when the response of lying down (Link 1) is immediately followed by the stimulus situation for Link 2, which is the "lying down" situation.

3. When the two previous conditions are fully met, it appears that the acquisition of a chain is not a gradual process, but one which occurs on a single occasion. (Of course, if these prior conditions are *not* met, the occasion may have to be repeated for the purpose of establishing the links themselves. This would be the case, for example, if the child had not fully learned the $Ss \rightarrow R$ connection of saying "Doll.")

Learning Type 4: Verbal Association

In view of the amount of attention devoted to the learning of verbal associates and nonsense lists since the time of Ebbinghaus, it would be unusual if we did not find a place for this type in a list of learning varieties. However, verbal association might well be classified as only a subvariety of chaining. There are short chains, such as that of a single pair of associates, like GUK-RIV, and long ones like 10- or 12-syllable lists that have been so extensively studied. But because these chains are verbal, and therefore exploit the remarkable versatility of human processes, verbal association has some unique characteristics. For this reason, it is described here as a fourth form of learning, to be established by a distinguishable set of learning conditions. It should be possible to discern what kind of chaining is involved, and what the links are.

One learns to translate an English word into a foreign one by acquiring a chain. For example, the French word for "match" is "alumette." In order to learn this equivalent most expeditiously, something like the following set of events occurs. One examines the combination "match-alumette" and notices that something already known connects the two: the syllable "lum" which occurs in the word "illuminate." One then runs through the sequence, not necessarily out loud, "a match illuminates; lum: alumette." For many people, the chain is most readily established by means of an image of a match bursting into flame, so that the entire chain would be something like this:

$$Ss \xrightarrow[\text{``match''}]{} r \bigtriangledown \sim s \bigtriangledown \longrightarrow r_{\text{illuminate}} \overset{\sim}{\text{lum}} \overset{s}{\longrightarrow} R_{\text{``alumette''}}$$

The internal parts of this chain, denoted by s's and r's, are likely to be highly individualistic, since they depend on the previous learning history of the individual. Thus, if a person doesn't know the word "illuminate," it is obvious that the chain he acquires must be an entirely different one. It may be a longer one, or a shorter one; and it may involve a visual image, or some other kind of internal representation.

The conditions for the learning of verbal chains of this sort would appear to be as follows:

1. An $Ss \rightarrow R$ connection must have been previously learned that associates the word "match" with the image of a burning match (or with whatever other internal response may be employed). In simple terms, the individual must "know what a match is."

2. Another $Ss \rightarrow R$ connection must have been learned that enables the individual to associate the key syllable *lum* with the response "alumette." In other words, response differentiation must have previously taken place. The individual must know how to say the word "alumette" with sufficient accuracy to be considered correct (whether or not he sounds like a native Frenchman).

3. A "coding connection" must also be available, that is, must previously have been learned, if the chain is to be established with ease. In this case, the code is represented by the association of the image of the flaming match and the word "illumination." As we have said, the selection of this code by the individual depends on his own previous history. A highly verbal person may have many codes available, whereas an individual who ranks low in such ability may have very few.

Probably, the code we have depicted here would serve adequately for a large portion of the adult population.

4. The chain must be "reeled off" in a sequence, so that each $Ss \rightarrow R$ is contiguous in time with the next; in other words, contiguity is necessary for learning. Under these circumstances, the chain, like other learned behavior chains, is probably acquired on a single occasion.

It is noteworthy that a young child may have few "codes" available to him. Thus he may have considerably greater difficulty in acquiring a verbal chain of this sort. If the words "match" and "alumette" are repeated together frequently enough, the child eventually will find a code that will enable him to learn the chain. If one wants him to learn it rapidly, however, presenting a distinctive code (perhaps in a picture) would seem to be an important aspect of the process. The efficacy of using pictures in the learning of foreign word meanings is well known.

What about the nonsense chains that connect the stimulus GUK with the response "riv"? There is no particular reason to believe that these require any different conditions for learning. The optimal conditions may, however, be more difficult to achieve. For example, $Ss \rightarrow R$ learning must be established, both for the stimulus member GUK and for the response member RIV. Then a coding connection must be selected, and there may be only a few available to the individual, depending on the "meaningfulness" (to him) of these syllables. He may have to resort to a somewhat lengthy chain, as would be the case if he said to himself "GUK suggests gook (engine sludge), which may affect the engine when you 'rev' it, or RIV."

Learning Type 5: Multiple Discrimination

Considered as isolated acts, the learning of single $Ss \rightarrow R$'s and the learning of chains of $Ss \rightarrow R$'s represent fairly simple events. Each seems easy, and in fact *is* easy, just so long as each instance of learning is carefully distinguished and insulated from other instances that may tend to occur at the same time, or from other instances of a similar sort occurring at different times in the same individual. But we also know that, practically speaking, making a permanent change in behavior by means of learning is not always so easy. The reason at once comes to mind: people readily *forget* what they have learned. The marvelous plasticity that characterizes the nervous system and makes possible these fundamental varieties of modification is counterbalanced by another characteristic: what has been learned and stored is readily weakened or obliterated by other activities.

When an individual acquires a chain that makes it possible for him to say "alumette" to "match," and then goes on to learn to say "fromage" for "cheese," he may by so doing weaken the first chain; he may forget the French word for "match." By experiment, it may also be shown that it is harder for him to remember "fromage" (for "cheese") than it would have been had he not first learned alumette for "match." If he tries to learn four French words at once, rather than two, this will be more than twice as difficult; six at once will be more than three times as difficult; and so on. Obviously, some new kind of process has entered the picture. Short chains are easy to learn but hard to retain. Increasing the number to be learned does not change the basic nature of the learning process, but it highlights the effects of another process—*forgetting*.

Many an American boy undertakes to learn to identify by name all the new models of automobiles produced in this country in any given year. He does not learn this in school, but it is surely as marvelous an accomplishment as many that do take place there. Each year there are new model names, as well as old ones, and these in turn are but subordinate categories of larger classes of names for major automobile manufacturers. Within a few weeks after all the new cars appear, a boy may be able to identify correctly the scores of new models that are adding to road congestion, as well as the ones he learned last year or the year before. His father, in contrast, may never get them straightened out—he has given this up long ago.

What the boy has acquired is a set of *multiple discriminations*. Each single identifying connection he learns is, of course, a chain. As a stimulus, each automobile must be discriminated from other stimuli, like trucks and buses, as the initial $Ss \rightarrow R$ connection in the chain that has as its terminal response the model name. As an individual chain, each is learned rather easily. But then something else is added to the situation. Each individual model, with its distinctive appearance, must be connected with its own model name and *with no other*. How is this ever possible, thinks the adult, when they all look so much alike anyhow? It is possible because they are in fact physically different, and they can be identified when one learns to make different responses to these physical differences. Twin headlights and a vertical grill, with chrome trim around the windshield, may constitute a stimulus combination that can be invariably connected with a single model name. A distinctive body shape may serve to identify another; and so on. In order to acquire multiple discriminations that identify all of them, the individual must first acquire a distinctive set of $Ss \rightarrow R$'s that differentiate the stimuli and set off the chains leading to the responses that are the model names.

But if one starts out to acquire these multiple identifications, it takes longer than one would expect from a simple summing of the occasions for the chaining of each identification. As we have said, the reason is that some get forgotten and have to be reestablished. The new chains *interfere* with the retention of those previously learned, and vice versa. The phenomenon of *interference,* which presumably is the basic mechanism for forgetting, is therefore a prominent characteristic of the learning of multiple discriminations. In fact, it may be said that the question of how to arrange the conditions for the learning of multiple discriminations becomes mainly a matter of reducing or preventing interference.

In brief, then, the conditions for learning multiple discriminations are as follows:

1. Individual chains connecting each distinctive stimulus with each identifying response must be learned. This, of course, implies that the individual $Ss \rightarrow R$'s that differentiate the *stimuli* must have been previously learned, as well as the *response* names themselves. (For American cars, these are usually chosen to be both highly familiar and distinctive.)

2. In order to ensure retention, measures must be taken to reduce interference. A number of methods may be employed. . . . Generally speaking, they have the purpose of making the stimuli as highly distinctive as possible. (A highly distinctive appearance of a car model virtually ensures that its name will be easily remembered.)

Multiple discrimination is the type of learning the teacher undertakes in order to be able to call each of her pupils by his correct name. It is the type that applies when the student learns to distinguish plants, animals, chemical elements, rocks, and to call them correctly by their individual names. In the young child, it is the kind of learning that happens in learning colors, shapes, common objects, letters and words, numerals and operation symbols. In the student of foreign languages, it occurs in the distinguishing of new words, as the sounds of *fin, femme,* and *faim,* as well as other less confusable ones (less subject to interference). It is, therefore, a kind of learning that is remarkably prevalent in all of formal education.

Is multiple discrimination "rote learning"? Perhaps yes, perhaps no; it depends on what is meant. As we have pointed out, the prominent characteristic of multiple discrimination is not the acquiring of new entities as such, since these are simply chains, each of which is readily acquired in isolation. Insofar as "rote" implies "committed to memory," the emphasis is surely correct: the important factor in multiple discrimination is the interference that must be overcome if retention is to be assured. But if "rote" implies repeated practice as an optimal method of learning, then this is not necessarily the case. . . .

Learning Type 6: Concept Learning

Attention now turns to a kind of learning that appears to be critically dependent on internal neural processes of *representation* for its very existence. In man, this function is served by language. Although a number of animals have been shown to possess the capacity to make internal representations of their environments, present evidence suggests that this capacity is extremely limited even in the higher apes. Human beings, in contrast, employ this capacity freely and prodigally; they are highly inclined to internalize their environment, to "manipulate" it symbolically, to think about it in endless ways.

Learning a concept means learning to respond to stimuli in terms of abstracted properties like "color," "shape," "position," "number," as opposed to concrete physical properties like specific wavelengths or particular intensities. A child may learn to call a two-inch cube a "block," and also to apply this name to other objects that differ from it somewhat in size and shape. Later, he learns the concept *cube,* and by so doing is able to identify a class of objects that differ from each other physically in infinite ways. A cube may be represented concretely by objects made of wood, glass, wire, or almost any material; the object may be of any color or texture, and of any size. Considering this great variety of physical stimulation that may correctly be identified as "cubical," perhaps it is not surprising that it takes some very precise language on the part of geometers to define what is meant by "cube." But, of course, a person does not have to understand such a definition in order to identify correctly a cube under most ordinary conditions of his existence. Except for some very special purposes of mathematical theory, an individual identifies a cube "intuitively," that is, on the basis of an internalized representation that does not employ the words of the geometer's definition. Whatever the process may be, there can be little doubt that a concept like cube is learned, and that its possession enables the individual to classify objects of widely differing physical appearance. His behavior comes to be controlled, not by particular stimuli that can be identified in specific physical terms, but by abstract properties of such stimuli.

As an example, we may consider how a child learns the concept *middle*. Initially, he may have been presented with a set of blocks arranged like this: ☐   ☐   ☐ . If previous $Ss \rightarrow R$ learning has enabled him to receive reinforcement for a request such as "Give me a block," he can then readily learn the simple chain of picking up the middle block when his parent says "Give me the *middle* one." Similar chains can then be established with other objects, such as balls arranged in the same configuration ◯ ◯ ◯ , or sticks | | | . One might want to make certain that the chains were generalizable over a range of separations, like | | | and | | | . In other words, the deliberate attempt is made to establish a number of chains applicable to a *variety* of specific physical configurations. Continuing this effort, the spoken word "middle" might be applied to various other arrangements of objects, including such situations as these:

By this means, the child comes to respond correctly to middle as a concept meaning "an object between two others." (Other meanings of "middle," such as the "center of an area," are, of course, different, but may be similarly learned.)

How does one know whether the child has in fact learned the concept middle? The crucial test is whether he will be able to respond correctly, not by chance, to some new configurations of objects he has not previously used in the course of learning. For example, these might be:

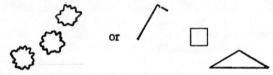

If the child can now respond properly when told "Give me the middle one," it may be concluded that he has learned a concept, and that his behavior is *not* controlled by specific stimuli.

Of course, it should be understood that a concept, once acquired, may also be associated by chaining with responses other than lifting. The response may be pointing, or pushing. No new events are involved in such instances; they are simply further examples of chain learning. It may also be noted that an essential part of the chain is the internal response "middle," which is the internal representation that permits discrimination of the correct final response from any incorrect response. This representational $s \rightarrow r$ may be spoken out loud as an overt verbal response "middle," or it may not be. In any case it functions as a word does. The individual must have a name for what he is about to select, even if the "name" is no more than an "ugh!" It is this name that selects and directs the final response, in the face of the varying stimulus situations encountered. A resemblance may also be noted between this naming response and the coding response that occurs in verbal association. Again, a similar function is being served.

The conditions for concept learning that are apparent in the preceding example are as follows:

1. The stimulus portion of the chain, by means of which a *middle block* is differentiated from the two others in the set, must have been previously learned. Likewise, the internal coding portion of the chain must have been previously acquired, which enables the individual to verbalize the word "middle" (or some other word serving the same function). And the response portion must also be available)saying "Middle," or pointing to it).

2. A *variety* of stimulus situations must be presented incorporating the conceptual property to be learned, in order that this property can become discriminated in its internally represented form.

3. Because of the necessity for this process of discrimination in a variety of different stimulus situations, the learning of a brand-new concept may in some circumstances be a *gradual* process.

One may expect that a child acquires many important concepts by means of this procedure. This is how he first learns what a circle is, or a square, or, for that matter, a cat (the class "cat" as opposed to a particular animal), or a tree. Presumably, he learns concepts like *up* and *down* this way; *near* and *far; right* and *left; larger* and *smaller; outside* and *inside;* and many others. Initially, the mistakes he makes, as in correctly identifying a saucer as circular, but not a dime, result from his having responded to an inadequate variety of stimulus situations. The concept may become more nearly adequate when he has a greater number of experiences in the course of everyday existence. Alternatively, it may become equally comprehensive as a result of a deliberate instructional process carried out over a much shorter period of time.

It is quite important to note, however, that adults do not always, or even frequently, learn new concepts in the manner previously described. Adults can take shortcuts because they have a greater fund of language. If an adult does not happen to know what *middle* means, he can learn it by acquiring a chain linking this word with another concept he already knows, such as *in between.* Suppose an adult were presented with the situation of three cubes in a row: ☐ ☐ ☐ . He is told "Give me the middle one." Now suppose he responds incorrectly. Assume also that he possesses the concept *in between* from previous learning. In order to learn this new and strange concept *middle,* it is simply necessary for another person to say, "Middle means in between." As a result of this communication, the adult acquires a verbal chain (type 4 learning) that becomes a part of the longer chain leading to the correct response.

This example of adult concept acquisition is instructive for two reasons. First, it illustrates how *verbal instruction* can function to remove the necessity for the gradual process of experience with a variety of stimulus situations necessary for concept learning in the child. By means of verbal chaining, an adult can acquire a new concept (at least a fairly adequate one) in a single trial. Second, the example emphasizes the difference between the genuine learning of a truly novel concept, as in a child, and the verbal generalization of an already learned concept, as in an adult. In some experimental studies, the second type of event has been identified as concept learning, whereas it might better be called *concept generalizing* or *concept using,* and distinguished from the concept learning that may be observed most clearly in the child.

Learning Type 7: Principle Learning

As an approach is made to a description of the kind of learning that can be identified as most broadly applicable to the content of formal education, the job grows somewhat easier. The most difficult kinds of learning to describe are the simplest kinds, perhaps because it is necessary to be so careful to keep them simple. A kind of learning like *principle learning* has some well-known conditions for its establishment, which all of us employ very frequently. Continuation of the basic theme of the argument—that different varieties of learning can be distinguished by the conditions required to bring them about—leads to the view that principle learning is not only a highly familiar but also a readily understandable kind of learning. As is true with certain other forms, the most important conditions are the prerequisites for learning.

Principle learning may be exemplified by the acquisition of the "idea" contained in such propositions as "gases expand when heated"; "the pronoun *each* takes a singular verb"; "salt is composed of the elements Na and Cl"; "$xa + xb = x(a + b)$"; "the definite article *die* goes with a feminine (German) noun"; and many, many others. Surely there can be little doubt that human beings must learn large numbers of such principles, some interconnected, some not, along the road to attaining the status of being considered educated adults.

Possibly the immediate demurral will be raised, "Why, these are simply verbal facts to be memorized!" Not so; and that is why the word "idea" has been employed to describe these principles, inexact as that word may be. From our previous discussion, it is apparent that each of these statements *can* be learned as a verbal chain. If we want to teach a five-year-old to memorize them in that way, this surely can be done. But the only kind of performance that would be possible following such learning would be something like this: Complete the statement: "The pronoun *each* takes _____ _____." Such a performance is by no means what is established when one has learned a principle. In fact, one may learn the principle about "each" and its verb without being able to verbalize such a statement at all. What *is* meant by learning such a principle is this capability: being able to use the singular form of the verb in a variety of sentences or clauses having *each* as a subject. If an individual is able to demonstrate this capability in a number of instances of representative sentences and clauses, one is justified in concluding that he has learned a principle.

In a formal sense, a principle is *a chain of two or more concepts.* Some would say, it is a *relationship* between concepts; but it seems preferable to state the nature of this relationship. The simplest type of principle may usually be cast in the form "If A, then B," as in the example "If a feminine noun, then the article *die*." The principles of chaining are essentially those that have been described for type 3 learning, except, of course, that concepts (internal representations) are being linked rather than simply $Ss \rightarrow R$'s. Accordingly, the conditions for the learning of principles appear to be these:

1. The concepts to be linked must have been previously learned. Thus, for an effective capability to be established in using singular verbs with

the pronoun *each,* the relevant concepts should be learned as prerequisites. The learner must have previously learned to respond to *each* as a concept, that is, as having a particular function in a sentence, and as different from other pronouns such as *they* or *all.* In addition, he must have previously learned *singular verb* as a concept differing from *singular noun,* and from *plural verb.* (Again, it is necessary to emphasize that we do not imply simply the acquiring of verbal chains by such learning, but the true establishment of concepts, as described under type 6 learning, leading to the capability of identifying the concepts correctly in a variety of stimulus examples.) In simple terms, when a principle is to be learned, the individual must already understand the concepts being chained.

2. Assuming that the first condition has been fulfilled, the process of chaining is a very simple matter. Usually, one simply states the principle verbally, as "The pronoun *each* takes a singular verb."

3. Under these circumstances, the learning of a principle takes place on a single occasion.

Some writers have suggested that principles need to be *discovered,* by which is meant that examples of the principle should be presented and the learner left to do his own chaining, without the help of verbal statements. The evidence is not particularly one-sided that this is the better method. After all, one should not forget that man is a verbal animal, and there are marvelous shortcuts in learning to be achieved by the use of language. The chances are that, provided condition 1 has been fully realized, telling the learner the principle verbally is much quicker than any other method, and may well be as effective (cf. Ausubel, 1963). But discovery of principles *can* certainly be made. What happens is that the individual selects his own idiosyncratic representations of the concepts to be chained.

The danger in using verbal statements as a kind of shortcut to learning lies mainly in the possibility that sheer *verbal chains* will be acquired rather than truly *conceptual chains.* All teachers are acquainted with the student who can *say* the principle without being able to *do* it. But this is an example of ineffective instruction. What it implies is that the student really has not previously acquired the concepts (condition 1 above); therefore, he reacts to the verbal statement as a verbal chain to be memorized. If one makes certain that the concept *each* is known, and the concept *singular verb,* the statement "*each* takes a singular verb" should in fact represent an optimal condition for learning. Of course, to test this, one must determine whether the learner can use the principle.

Learning Type 8: Problem Solving

Once he has acquired some principles, the human being can use them for many purposes in dealing with and controlling his environment. He can also do something else that is most important: he can *think.* Basically, this means he is able to combine the principles he has already learned into a great variety of novel higher-order principles. He may do this by stimulating himself, in part, and also by responding to various forms of stimulation from his environment. By means of the process of combining old principles into new ones, he *solves problems* that are new to him, and thus acquires still a greater store of new capabilities. The problems he solves are new to him, but they may not be new at all to other people. Of course, every so often a scientist or other creative person may arrive at a problem solution that is novel to society in general.

Problem solving, by which is meant "thinking out" a new principle that combines previously learned principles, is a process that is very familiar to most adults. There is nothing very "special" about such events, since they are likely to occur with a frequency of ten or twenty times daily in the life of an average man. When a driver maps his route through traffic (as opposed to simply being swept along by it) he is solving a problem. When a man replans his luncheon schedule as a result of a new appointment, he is solving a problem. When a housewife decides to shop selectively for certain items on the basis of differential prices, she is solving a problem. These everyday examples bear a close formal resemblance to the problems that are solved by students in composing reports and themes, in marshaling arguments to present a point of view, in performing laboratory experiments, and in reading a properly written textbook.

Suppose that a student of physics encounters this situation in reading his textbook:

Power is defined as the work done divided by the time during which this work is done. That is,

$$\text{Power} = \frac{\text{Work}}{t}$$

Suppose we have a body doing work on another body, acting with force F and velocity v. Can you show that the power delivered to a body is the product of the force acting on it and its velocity?

The solution of such a problem requires thinking, which might go somewhat as follows. First, the student formulates to himself the fact that he wants an expression relating power to force (F) and velocity (v). He then recalls that work is related to force and the distance through which it moves, that is,

$$\text{Power} = \frac{\text{Work}}{t} = \frac{Fs}{t}$$

and he is therefore able to state that

$$\text{Power} = Fv$$

and to recognize this as the formulation he is seeking. Involved in these steps, of course, is the use of other principles he has previously learned, pertaining to the substitution of equivalent quantities in equations.

When these steps have been accomplished, it is reasonable to suppose that the student has learned something new. He has not learned simply that Power = Fv, because he could readily have learned that as a verbal sequence (type 4). Rather, he has learned how to demonstrate in concrete terms the relationship of power to force and velocity, beginning with the definition that power is work done per unit time. In other words, he has learned how to *use* this definition, and how to generalize it to some considerable degree to situations in which a force acts against a body with a known velocity. This act of problem solving, then, has resulted in some very substantial learning. The change in the individual's capability is just as clear and unambiguous as it is in any other variety of learning.

A number of conditions can be identified as apparently essential for this act of learning:

1. The learner must be able to identify the essential features of the response that will be the solution, *before* he arrives at it. (Some writers on problem solving have said that the learner must have a *goal*.) This particular condition appears to be important because of the lengthy chains involved, and the steplike character of the problem-solving act. In the present example, the student knows he needs an equation relating power, force, and velocity. Presumably, the function of this identification is to provide directedness to thinking. By checking his successive responses against those of this simple chain, each step of the way, the learner is able to "keep on the track," and also to "know when to stop."

2. Relevant principles, which have previously been learned, are recalled. In the example given, the learner recalls specifically the principles 1) Work = $F \cdot s$, and 2) $v = s/t$, as well as mathematical principles such as $(ab)/c = a(b/c)$, and if $a = b$ *and* $b = c$, then $a = c$. (It should be noted that in all probability, these mathematical principles would be highly recallable for the usual student of physics, because they have already been so frequently used.)

3. The recalled principles are combined so that a new principle emerges and is learned. It must be admitted here that little is known about the nature of this "combining" event, and it cannot now be described with any degree of completeness. Simply writing down in a sequence the "logical steps" followed in the thinking does not answer this question. These steps are the intermediate responses made by the learner, that is, they are the *outcomes* of his thinking. But they provide few clues as to the nature of the combining process itself.

4. The individual steps involved in problem solving may be many, and therefore the entire act may take some time. Nevertheless, it seems evident that the solution is arrived at "suddenly," in a "flash of insight." Repetition has little to do with it. Nor is repetition a very powerful factor in the prevention of interference, or forgetting, as is the case with multiple discrimination learning. A "higher-order" principle resulting from an act of thinking appears to be remarkably resistant to forgetting.

Problem solving results in the acquisition of new ideas that multiply the applicability of principles previously learned. As is the case with other forms of learning, its occurrence is founded on these previously learned capabilities; it does not take place "in a vacuum," devoid of content knowledge. The major condition for encouraging the learner to think is to be sure he already has something to think about. Learning by problem solving leads to new capabilities for further thinking. Included among these are not only the "higher-order" principles we have emphasized here, but also "sets" and "strategies" that serve to determine the direction of thinking and therefore its productiveness.

References

Ausubel, David P. *Psychology of Meaningful Verbal Learning.* New York: Greene & Stratton, 1963.

Berlyne, D. E. *Structure and Direction in Thinking.* New York: John Wiley & Sons, 1965.

Bourne, L. E., Jr. *Human Conceptual Behavior.* Boston: Allyn and Bacon, 1966.

Bruner, J. S., Goodnow, J. J., and Austin, G. A. *A Study of Thinking.* New York: John Wiley & Sons, 1965.

Ferster, C. B. and Skinner, B. F. *Schedules of Reinforcement.* New York: Appleton-Century-Crofts, 1957.

Gagné, R. M. *The Conditions of Learning.* New York: Holt, Rinehart and Winston, 1965, 1970.

Gagné, R. M. "Problem Solving." In *Categories of Human Learning,* A. W. Melton, ed. New York: Academic Press, 1964.

Gilbert, T. F. "Mathetics: The Technology of Education," *J. Mathetics,* 1 (1962): 7-73.

Guthrie, E. R. *The Psychology of Learning.* New York: Harper & Row Publishers, 1935.

Kendler, H. H. "The Concept of the Concept." In *Categories of Human Learning,* A. W. Melton, ed. New York: Academic Press, 1964.

Kimble, G. A. *Hilgard and Marquis "Conditioning and Learning."* New York: Appleton-Century-Crofts, 1961.

Mowrer, O. H. *Learning Theory and Behavior.* New York: John Wiley & Sons, 1960.

Mowrer, O. H. *Learning Theory and the Symbolic Processes.* New York: John Wiley & Sons, 1960.

Northrop, R. S. C. *The Logic of the Sciences and the Humanities.* New York: Macmillan, 1947.

Pavlov, I. P. *Conditioned Reflexes.* Trans. G. V. Anrep. London: Oxford University Press, 1927. (Also available in paperback; New York: Dover.)

Postman, L. "The Present Status of Interference Theory." In *Verbal Learning and Verbal Behavior.* C. N. Cofer, ed. New York: McGraw-Hill, 1961.

Skinner, B. F. *The Behavior of Organisms: an Experimental Analysis.* New York: Appleton-Century-Crofts, 1938.

Thorndike, E. L. "Animal Intelligence: An Experimental Study of the Associative Processes in Animals." *Psychol. Rev. Monogr. Suppl.,* 1898, 2, No. 4 (Whole No. 8).

Underwood, B. J. "The Representativeness of Rote Verbal Learning." In *Categories of Human Learning,* A. W. Melton, ed. New York: Academic Press, 1964.

Wertheimer, M. *Productive Thinking.* New York: Harper & Row Publishers, 1945.

Exercises: Varieties of Learning

1. Give an example from the classroom of signal learning. How does stimulus-response learning differ from signal learning? Compare the learned responses. How are they different? How about the stimuli? Think of an example of stimulus-response learning with children in a classroom. Does your example from the classroom of signal learning still seem correct?

2. Can signal learning or stimulus-response learning be verbal? Why? Justify?

3. From the classroom, give an example of an undesirable chain, the corresponding desirable chain, and a strategy for teaching the desirable chain as a replacement for the undesirable chain.

4. If a child memorizes pi to thirty-five places, i.e., can say "3.14159265358793238462643383827950288," we may classify this as verbal association. It is probably a sequence of verbal stimuli and responses which were "rattled off" by the student "without thinking." Give two other examples of verbal association with children in the classroom. Is verbal association the first learning type which involves language?

5. What kinds of chains must be formed by a second grader in learning to use a compass to draw a circle? What chains are formed in using a protractor to measure the size of an angle?

6. Perhaps learning to recognize people and remember their names is an example of multiple discrimination. How many of us have failed in this task when we have come up behind someone who appears to be a friend only to discover that it is a complete stranger we have "recognized." Give an example of multiple discrimination from the classroom. Is Gagné correct in claiming that verbal associations are a necessary prerequisite to multiple discriminations? Support your answer using your example of multiple discrimination.

7. What kinds of multiple discriminations are necessary in evaluating the expression

$$x = \frac{3 + 2\,(8 \div 4) \cdot 7 - 1}{9 - 2 + 1}$$

8. What is a concept? Give an example from the classroom and show that it satisfies Gagné's definition.

9. Is the number fact "3 + 4 = 7" a concept, a stimulus-response pairing, or a verbal chaining task?

10. Why is it necessary to provide new instances to the student when you are testing for the learning of a concept? If you use familiar instances, what kind of learning could you be measuring? Give an example from the classroom where it would be easy to confuse verbal association with concept learning. What can be done to prevent this confusion?

11. Give several examples of principle learning from the classroom. How can you distinguish, on a test, the difference between verbal association and principle learning? How do you think the proof of a mathematical principle fits into principle learning? Give an example.

12. Are the following statements concepts or principles:
    a. Addition is commutative.
    b. $\neq$ means not equal.
    c. A triangle has three sides.
    d. The measure of a right angle is 90 degrees.
    e. Eight is an even number.
    f. All prime numbers greater than two are odd.

13. The phenomenon of "interference" is discussed by Gagné as a mechanism of forgetting. Cite an example of mathematics learning that you feel is particularly susceptible to "interference." Discuss how you might minimize interference in this case.

14. To ask a student to factor $x^2 + 5x + 6$ is not a problem if the student has seen or worked examples of the form:

$$x^2 + (a + b) x + ab = (x + a) (x + b)$$

On the other hand, if the learner has had no experience with factoring trinomials and the learner must combine

$$a (b + c) = ab + ac$$
$$a + (b + c) = (a + b) + c$$
$$ab = ba$$

in a new manner not previously known, then factoring $x^2 + 5x + 6$ may indeed be problem solving.

Choose a mathematics textbook and identify some examples of problems (Gagné type). Show by describing the material presented to the student before each problem that what you have chosen will require problem solving. (Warning: Mathematics textbooks do not often ask for problem-solving behavior).

15. Construct what you consider to be a first-rate, mathematical example of each of the eight learning types.

16. Select a verbal problem from ninth grade algebra. Work this problem in detail then identify behavioral objectives related to this specific problem that reflect Gagné's eight types of learning.

17. Gagné made the claim that each learning type is a prerequisite for all the learning types which follow in the sequence:

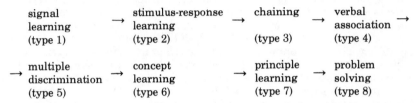

signal learning (type 1) → stimulus-response learning (type 2) → chaining (type 3) → verbal association (type 4) →

→ multiple discrimination (type 5) → concept learning (type 6) → principle learning (type 7) → problem solving (type 8)

Often learning theories are logically constructed to favor themselves. That is, it appears at first that there is no possible "other way" to do things. Has Gagné set it up that way, or is it really the case that each learning type is a prerequisite for the others that follow it? Can most human learning be classified as one of the eight learning types given by Gagné?

Issues Current in Educational Psychology

Lee J. Cronbach

I was originally asked to describe the significance of the new curricular movements, particularly in mathematics, for educational psychology. As I wrote, however, I found myself approaching the theme from another angle, asking, what are the big open questions in educational psychology? The new curricula have had their part in opening up these questions. The reactionary critics of education and the new adventurers both talk of mental discipline, and psychologists are being forced to reconsider whether the concept has legitimate meaning or is only a haunting echo from an ancient fallacy. No one trained as a teacher to connect abstract knowledge with the "real problem" on which it bears can, without a shock, hear representatives of the Physical Science Study Committee state their conviction that the high-school physics course has no proper place for the mention of refrigerators and household wiring. The intelligence of the proponents and the obvious quality of their work requires us to give a respectful hearing when they challenge articles of the educational psychologist's faith. Frequent interchanges with curriculum developers in the last three years have helped me face psychological issues more squarely. Here I am going to discuss those issues, however, more than the curricula themselves.

Educational Psychology: Dark Ages and Renaissance

From 1940 to 1954 the educational psychology taught to teachers and considered in curriculum planning was almost static.[1] While not much new knowledge has been consolidated since 1954, a combination of influences—the new curricula, for one—has brought us a fresh perspective. The dates 1940 and 1954 are not precisely landmarks in educational psychology; 1940 marks only when I happened to complete my own doctoral work, and 1954 the appearance of my textbook in

---

Lee J. Cronbach, "Issues Current in Educational Psychology," *Mathematical Learning,* Morrisett and Vinsonhaler, eds. *Monographs of the Society for Research in Child Development,* 1965, **30**, Ser. #99, pp. 109-126. Copyright © 1965 by The Society for Research in Child Development, Inc. Reprinted by permission.

[1]It is not that educational psychologists, as an isolated group, were failing to keep up with their science; the pertinent science itself was, on the surface, static. Neal Miller, an outstanding student of learning in the laboratory, was prevailed upon in 1954 to produce a book (Miller, 1957) spelling out the implications of learning theory for educational films and other "graphic communications." The concepts and principles that he used (very successfully) were those of a 1941 book (Miller & Dollard, 1941) only fleshed out by experience from wartime military applications.

educational psychology (Cronbach, 1954), but, give or take a couple of years, this does mark off a significant period: the period in which commerce between academic psychology and educational psychology was cut off by a tacit embargo. Persons concerned with education found no nourishment in the systematic Hullian studies of T mazes and eyeblink conditioning that began to dominate experimental psychology in the late 1930's. Experimental psychologists were repelled by the educator's insistence on talking about "the whole child" in "real-life situations"—both being prescientific or even antiscientific phrases antithetical to the analytic, formal style that, for a time, was the ideal of American behavior theory. My colleague, J. McV. Hunt, tells me that the psychologist's aversion to educational entanglements was manifest even early in the century. Thorndike was respected in spite of rather than because of his concern with education. One eminent experimenter is reported as turning down a position, in 1921, that would have doubled his salary because it was connected with a School of Education. And Hunt recalls scornful discussion among the eminent Cornell staff of the 1930's about "the drag of education upon psychology with its demand for little steps for little feet."

The educational psychologist, during the years of estrangement, felt that the teachers he taught needed little from theoretical or laboratory psychology.[2] He worked hard and ingeniously to improve teachers' skills in analyzing the individual learner's aptitudes, emotional difficulties, and educational deficiencies. He extracted important educational implications from the burgeoning clinical and social psychologies. But he stopped doing research on intellectual learning or on "psychology of the school subjects," in part because these were not matters of live interest in educational circles. The wave of research into educational learning broke and drained away. Its last major manifestation, I suggest, was the 1936 book entitled *Education as the Cultivation of Higher Mental Processes,* by Judd (1936).[3] This book, with its out-of-fashion title and message, received little attention. Subsequent to 1940, only a few notable reports by such workers as Brownell dealt with intellectual learning. (Brownell concentrated on the development, through meaningful presentation and insightful practice, of skill in computation.)

We can pass over the 1940's, noting only that young experimenters blooded in the aviation research laboratories of the 1940's and early 1950's are taking over the front line in today's educational psychology. This is a new breed of educational psychologist, firmly grounded in theoretical psychology and technical research method, who did not, like earlier workers in the field, come up through public school teaching. (A representative list of these newcomers would include Crowder, Fleishman, Gagné, Glaser, Lumsdaine, and Stolurow.) These men might

---

[2]It was in 1937 that Geldard, an experimental psychologist, told teachers that "educational practices will have to be guided for some time to come by rules of thumb. . . . This is not to say that nothing of interest or importance has been done [concerning] learning. Our happy plight at present is that we have at hand far more laws of learning than can possibly be true" (Geldard, 1937). Such a view is occassionally echoed even today (Spence, 1959).

[3]About one-third of the book deals with mathematics. The argument closely matches Bruner's (1957), when he speaks of "the structure of knowledge," although it makes much less place for intuition than he.

well have stayed in military training research, save for the coincidence that such programs were cut back just as national interest in education revived.

Today's work on intellectual processes is not merely a reaction to Russian spacemanship, although some pursestrings were loosened by it. Nor is it primarily a response to the curriculum programs. Movements within psychology emerged at a strategic time to offer new prototheories and new questions. Skinner's well-known 1954 paper on teaching machines was the by-product of 20 years of operant conditioning research. Perhaps more important as a forerunner of the new trend was Osgood's 1953 book, in which he elaborated a "mediation theory" to account for transfer of learning and thus made room within formal behavior theory for the internal verbal and conceptual processes that early behaviorism had tried to ignore. Subsequent work by other neo-Hullians (Berlyne, Lawrence, and Kendler, for example) brought problems into the laboratory that are not far removed from those of education; at the same time, the Russians, in a similar way, stretched Pavlovian theory to accommodate symbolic processes (Simon, 1957; Brozek, 1962). Two distinctive influences remain to be mentioned: Harlow and Piaget.

Harlow's (1949) striking paper on learning sets opened Americans' eyes to the fact that, in a certain reasonable interpretation of the phrase, intelligence is learned and, furthermore, that this learning can be cultivated under laboratory, experimental conditions. If you work with small rodents, you rear a litter in innocence, run them through your mazes for a week or so at the proper age, and then destroy them. But the primates with which Harlow worked are too expensive to use once and destroy. So Harlow had to study how one learning experience modifies the organism's approach to the next, and how these effects culminate over a long span.[4] Thus, the psychologist was forced to think like an educator.

The monkey acquired a learning set—presumably an awareness that there is a right answer to be looked for. Just such a learning set was formed, after countless trials, at that famous moment when Helen Keller learned that the tapping on her hand was connected with the water flowing over it, and so achieved the great insight that each thing has a name to be learned. In such teaching as David Page's, which seeks to convey to pupils that mathematics is a system of perceived regularities and that they can create mathematics rather than merely bow to it, the intended outcome is a learning set.

The notion that we learn to observe and to use information was not new; it was clearly present in Bartlett's concept of schemata and in Piaget's numerous observations of thinking in the child. But these transatlantic views, first encountered in the 1930's by Americans, did not

---

[4]Harlow confronted a monkey with three objects. On every trial a raisin was under the same object, say a blue cube. Only after many trials did the monkey always reach for the cube. Once this was mastered, the monkey had to solve a new problem with the same rule but a different object. Harlow found that in the course of 50 or so problems the monkey became efficient. On the first trial with new objects he had no recourse save trial and error. But the now-sophisticated monkey almost always made the right selection on the *second* trial. He had learned to extract information from experience.

fit into prevailing theories and methodologies; and so they were set aside until recent translations of the mature Piaget came to hand just as cognition came back in style.

There was no single reason for the revival of interest in cognition. Harlow's neat work contributed, and so did the engineering psychologist's flirtation with information theory. Investigators of human problem-solving, such as Bruner (1957), became concerned with the organization of ideas—a concept foreign to the old associative-learning theory. Among some specialists in education there was a match return to questions about intellect, of which the notable example was the *Taxonomy of Educational Objectives* (Bloom, 1956) prepared by leaders in measurement. This recognized that disciplined relations between ideas is a type of learning to be investigated and encouraged in school. As a model for examination construction it departed radically both from the sampling of fragmentary "facts" and the pragmatic testing of the student's ability to apply isolated principles to concrete cases.

What Educational Psychologists Have Believed

The issues I propose to discuss arise from re-examination of matters that were regarded for 20 years or so as settled, and, therefore, it will help to review those accepted beliefs. The chief "principles of learning" of the educational psychologist, vintage 1940, can be represented by the following statements. This list is not exhaustive, nor are the statements put in the sophisticated form that might have been in the professor's mind. More nearly, these are the simplfied views the educator remembers and which, therefore, influence him.

1. *Learning occurs through active practice.* This encouraged drill, but it has other implications when additional principles are brought into the picture.

2. *Pupils should not try tasks in which they are unlikely to succeed*—"for which they lack readiness." This led curriculum makers to postpone topics believed difficult.

3. *Transfer of a learned response is to be expected only if the later stimulus is much like that on which the person was trained.* Corollary: classroom tasks and problems should be lifelike. Here is the origin of the "social arithmetic" of which mathematicians have been so critical.

4. *A response that leads to a desired goal will be easier taught than one motivated only by external incentives and compulsions.* Corollary: develop the topics for study out of the interests of each particular class.

5. *Learning is shown to be "meaningful" if the pupil can use his knowledge in new situations, particularly concrete situations.* This is the principle of applicational transfer, emphasized in the 1930's by Pressey and Tyler as a counteraction to Thorndike's early emphasis on drill in situations closely similar to that to be encountered later.

6. *Factual learning or learning not clearly understood is quickly forgotten.*

7. *A well-understood verbal generalization is remembered, and aids in adaptation to new conditions.* This favors the teaching of abstractions, but in close connection with their concrete referents.

All these principles are sensible and supported by some research. They contradict alternative views once accepted in educational theory and practice, that is, they are not truisms. All these statements deserve continued consideration by educators. Yet each statement or its corollary is partly false, open to dispute, or seriously incomplete, in the light of current research. We shall have to reconsider the evidence on which the statements rest, the value judgments they conceal, and their theoretical underpinnings. Ultimately, we may hope to distil out their essential truth and define the conditions in which they apply, and also learn the conditions under which some other generalization holds. The ensuing sections will discuss only a selected few of the searching questions or criticisms now in the air.

I feel constrained to put in a paragraph in defense of myself and my colleagues. This paper, written with the aim of stimulating argument at a conference, is an overdramatic report on the ignorance of educational psychologists, something like that fourth-grade version of history in which everybody before Columbus believed the world to be flat. Critical thinking about these issues is nothing new. The wisest educational psychologists among my seniors—Gates and Brownell are worthy examples—remained thoroughly aware of and explicit about most of the points that I shall make, and of the sound ones that Bruner (1960) and Judd have made. The difficulty in educational psychology is that in speaking to teachers one propagandizes for some worthy change and so emphasizes certain aspects of one's theory to the neglect of others. In fighting a battle against classroom authoritarianism, for example, one does not pause to spell out the legitimate case for the teacher as authority. I should add that this paper confines itself to the doubtful matters in educational psychology: it says nothing about the observations and interpretations that are beyond suspicion. Educational psychology has much to tell teachers that is worth telling.

Prompting vs. Discovery

Of all the topics currently in the air, the sharpest split among psychologists is on the issue of prompting vs. discovery. We have, at the one extreme, the reinforcement theorists who follow Skinner's radical behaviorism. To them, a response is a response. Get clear in your mind what you want the pupil to *do* (mark that word well), construct a set of cues, or prompts, sufficient to provoke that response from him, and when he makes it, reinforce (reward) him at once. In time, he will have discriminated this from competing responses and will make it when only a few of the original cues are present. This theory is strongly reminiscent of the early Thorndike's which led teachers to use repetitious drill. Skinner's teaching machine, indeed, has been advocated primarily as a way of administering practice efficiently. Presentation of ready-made verbal generalizations is common in teaching of this type. Homme & Glaser (1961) advocate a "ruleg" system of programming in which you 1) give a verbal rule, 2) run application items to make sure the pupil knows what the words mean and how to use them, and 3) run verbal completion items to make sure the pupil has all the words in the right places. Other forms of teacher presentation and pupil practice have been the main standby in teaching science, English, and mathematics as well as in vocational training. The teacher who demonstrates an algorism—e.g., for division of decimals—first defines for himself the ideal overt response and then teaches directly to have the student execute it. In fancy language, he "exerts stimulus control to cause the pupil to emit the response."

The opposite extreme position holds that concepts, generalizations, and procedures ought to be created or discovered by the pupil. This position has been a dogma of the Illinois (UICSM) mathematics program and finds strong endorsement among some psychologists. Handing down a neatly packaged generalization, we are warned, stops the pupil from going through the preverbal adjustments of perceptions and trial actions that produce insight (Hendrix, 1961). Some persons (not in UICSM) argue that one should never lift the insight to the overt verbal level. Hendrix and Beberman, however, *lead the pupil* to construct the verbal statement after he has captured an idea in a preverbal way. In educational psychology, discovery and insight have been glorified since the work of Katona (1940) and other Gestalt psychologists in the 1930's, but we have never had a clear definition of the theoretical issues.

I am enthusiastic about teaching by discovery as a means of showing how knowledge originates. I do not want pupils to grow up slaves to formulas; I, like my colleagues in mathematics, want them to see knowledge as man-made and themselves as possible makers of it. I would like the girl to learn cooking in part by inventing new dishes. But once she has the basic concept of creativity in cooking I believe that she can follow a cookbook on most occasions without educational loss.

The questions are not about the motivational value of learning by discovery; it is its cognitive significance that concerns us. Must one discover a relation in order to understand it? Is discovery necessary to produce conviction rather than mere acquiescence? To begin with, let me try to skim off some of the irrelevancies and slogans that are tossed into the argument from both sides (Ausubel, 1961; Bruner, 1961; Hendrix, 1961),

1. "Discovery is democratic, didactic teaching is authoritarian."
   To be sure, the authoritarian is likely to be didactic. But reasoned presentation is consistent with democracy, most obviously so in mathematics, where dependence on authority is entirely unacceptable save in matters of convention.
2. "Discovered knowledge is meaningful, knowledge presented verbally is not."
   This is true only for stupid verbal presentation. Didactic teaching can be highly meaningful. That which is taught by discovery, moreover, is at best meaningful only to the student who discovers it, not to the many who fail to make the discovery.
3. "Discovery-before-words grounds the words in perceptions; giving words first does not."
   But words can certainly be presented in firm connection with their concrete referents.
4. "The student cannot discover Newtonian theory unless he is another Newton."
   This is no more than a debater's point. No teacher is truly teaching unless he is either arranging the conditions for discovery to occur, or explaining. A student, guided by a teacher who understands Newton, can reconstruct Newtonian principles.

The psychological truth has been obscured by partisan experimentation: technically unsound studies, studies that require "discovery" of arbitrary rules, studies that mislabel meaningful didactic teaching as "discovery," and so on. The fundamental questions are more subtle than is suggested by the proposition that learning by discovery is always best, or its opposite. The psychologist must identify the conditions under which induction or hypotheses-construction by the pupil is advisable, and he must learn the principles regulating the amount and timing of exposition by the teacher.

Ervin's (1960) recent study of third- and fourth-graders is an example of what is needed. She led pupils to discover the principle of reflection by means of experiments with a marble shot from a tube against a barrier. One group was given nonverbal aid in observing relevant facts: for example, the path of the marble was traced on the cloth and the equal angles were colored in. The other children were led by questions to work out the verbal principle from their observations. All instruction was individual. The most important result came from a transfer test in which a flashlight was to be aimed upward at a mirror with a back-to-front tilt so as to reflect on a target. Both groups improved, but there was no appreciable difference in their average scores. What is noteworthy is the result on the last, very difficult test item, in which the mirror was tipped sharply upward. On the earlier items, children could succeed by aiming about halfway between the vertical projection of the flashlight and that of the target onto the mirror, that is, by using an inexact, impressionistic principle. On the last item, the beam set in that manner would reflect onto the ceiling; only those who adjusted the angle of incidence could be right. Success on this problem was conspicuously greater among those who reached the correct verbal rule at the end of training, and this occurred most often in the verbally trained group. Moreover, the verbal training improved ability to formulate the rule for another problem. On the other hand, there were a fair number of pupils who attained the correct verbal rule on the marble problem and yet did not set the mirror properly. (Note that both groups here had guided discovery.)

In another well-designed study of discovery, Gagné & Brown (1961) prepared programs to teach high-school boys to develop a formula for the sum of a numerical series, e.g., arithmetic series. Instead of testing transfer by having subjects take the sums of other series of the same type, they tested *ability to construct new formulas* for series of unfamiliar types, e.g., of consecutive cubes. Three programs were prepared. One (*RE*) gave the formula—rule—for summing each training series and taught the pupil to apply it to examples. One demanded discovery (*D*) of the rule and gave a few hints as needed. The third broke the task into 40 steps of guided discovery (*GD*), each calling for analysis of a small bit of the series. All groups improved from one training series to another. On the test, they were asked to find the rule for a new series using hints as in the *D* training. Mean weighted time scores were: *RE,* 47; *D,* 28; *GD,* 23. Guided discovery was best. (It would be interesting to see comparable results on a program directly teaching rules for finding formulas, with examples. This would be rule-and-example teaching of a generalized procedure.)

We here observe, under experimental control, the teaching of a heuristic, a type of education not previously investigated formally. Even the most "meaningful" instruction in older mathematics curricula emphasized mastering certain theorems or models and applying them in appropriate specific situations. The new curricula, however, claim to be training the pupil to create new mathematical models or theorems, and Gagné is here showing how such learning can be demonstrated and investigated. The study also makes clear that the antithesis between discovery and programmed instruction is false. While Skinner does not favor the deliberate construction of a response by the subject, not all programmers are orthodox Skinnerians. Stolurow is directing an effort with UICSM to write a program that will elicit discovery of mathematical conclusions. This research will help to define the limits of programmed teaching.

It is premature to state firm general conclusions. Results will depend on the learner's maturity, his relevant concrete experience, and the symbolic systems he has at his command. The greater these are, the less I expect him to profit from the experience of discovery. We need to relate method to material to be mastered. Knowledge that can be verified experimentally or by its internal consistency is, I suspect, more appropriate for "discovery" than knowledge that is conventional (word meanings) or factual-descriptive. I think it pedantic to require children to "discover" that magnets attract iron, but they probably cannot comprehend the properties of a magnetic field save by exploratory investigation. We need experiments that carefully control the time allocated to discovery, to know how much slower that method is—for Gagné and Brown it was faster! And we need long extended, carefully observed educational studies to augment laboratory studies. I think we will find that a rich mixture of discovery-to-presentation is best to get the learner started, but that after he is well on the road, a leaner mixture will make for faster progress and greater economy.

Is Practice Really Necessary?

No principle is so central to American psychology as learning through action. We have an array of sound experiments showing that answering questions imbedded in an educational film increases learning, reciting lessons to oneself improves study efficiency, and, in skill learning, more practice makes more perfect. This has led to recommendations for increased recitation, increased drill, more frequent tests, laboratory courses in science, activity programs—the whole range of "learning by doing." To be sure, someone satirically pointed out the oddity that, according to the psychological literature, "German rats sit and think, while American rats rush about making trials and errors." And a few of us heard Stoddard's protest to educators that "we learn not by doing, but by thinking about what we are doing." Nonetheless, we have remained action-oriented, right down to the teaching machine of the moment.

The Bruner-Miller nonbehaviorist faction of Harvard psychologists drew my attention to the fact that one sometimes learns best by *not* doing. They have a little experiment (Bruner, Wallach, & Galanter, 1959) in which the subject is to predict whether the left or right light in a display will go on next. If a regular series such as *RRRLLLRRRL* . . . is presented, an S allowed to watch for 20 trials before starting to predict will do significantly better than the S who is required to guess from the outset. Shortly after hearing this, I became aware of a stream of negative results coming from trials of programmed instruction. In several experiments (Silverman & Alter, 1960; Goldbeck & Campbell, 1962;

Kieslar & McNeill, 1962; Krumboltz & Weisman, 1962) one group of students filled in the blanks in the statements of a programmed learning sequence and received immediate feedback of the correct answer; a second group read the same statements with the answers already filled in and underlined. The groups did equally well on posttests. This is a body blow to Skinner's rationale; it is a bit hard to see why the studies are reported so nonchalantly. Each study was done in a different laboratory; perhaps they are not impressive until one sees them all together. Explanations are seriously incomplete. George Miller suggests that, in information reception, guesses by the learner at the beginning introduce "noise" that impedes learning. This does not account for the results with teaching programs, for these are supposed to have an internal logic that eliminates noise.

I am inclined to the view that teaching machines are already obsolete. The machine rode in on the shoulders of reinforcement theory and automation, but Skinner's demand that the learner be led to the correct response on each trial quickly made the program central and the machine peripheral. The programming format has required unprecedented attention to the details of the stimulus series in teaching and to its internal logic.[5] The programmer is learning to write so that the learner always knows what he is about—an accomplishment beyond reach of the textbook writer, because he does not test comprehension after every sentence and so does not find out when he is confusing the pupil. But once the program approaches this perfection, the machine has done its part. Now the writing is so lucid and so well structured that the reader cannot escape the meaning even when it is presented in ordinary text form with the blanks filled in. Active-trial-and-reinforcement is most needed when lessons are nonsense, and become less and less necessary as presentations become clearer. If I am right, the machine will, in time, be relegated to its proper role as a proving ground for text materials in draft form and as a laboratory instrument helping us toward a science of clarity.

Meaning a Product of Cumulative Experience

I turn now to a study with broad implications. In one of the best-executed of all educational experiments, Brownell & Moser (1949) set out to determine once and for all whether subtraction should be taught in terms of borrowing (decomposition) or equal additions. They sought also to produce evidence that the results depended on the meaningfulness of instruction.

$$
\begin{array}{cc}
{}^{3}4{}^{1}2 & 4{}^{1}2 \\
\underline{2\ 7} & {}^{3}2\ 7 \\
1\ 5 & \underline{1\ 5}
\end{array}
$$

Decomposition	Equal addition

---

[5]There is an intriguing parallel between programmed presentation and Socratic teaching. In the latter, questions are asked to which the student will give one of a limited set of answers. As in a branching program, the answer identifies which question should be asked next. But responses are constructed by the learner whereas, in most programming, the responses are explained or demonstrated to him.

Four groups, several hundred pupils in each, were taught to subtract two-digit numbers:

1. *D-M*—borrowing rationalized.
2. *D-A*—borrowing as a straight-forward algorism. "Follow these steps."
3. *EA-M*—equal additions explained.
4. *EA-A*—equal additions as an algorism.

Groups *D-A* and *EA-A,* taught mechanically, were nearly equal in performance. The *D-M* group did better than the other three on direct tests and very much better on subtraction with three-digit numbers—which had not been directly taught. The *EA-M* group was intermediate in performance; none of the explanations given was well understood by the pupils. The conclusion was: teach subtraction by the *D* method for which an intuitively clear explanation can be given.

Now I am told that this recommendation is wrong. I have seen no evidence to support the claim, but the argument is plausible and merits a formal test. Richard Griggs, working in the Illinois Arithmetic Project, developed a new method of explaining *EA*. The child is asked to count steps along the number line: 27 + 5 steps gets us to 32 and one giant step of 10 units get us to 42: 5 + *10* = 15. This is readily comprehended and boils down to *EA*. Whether *this EA-M* method would work better or worse than Brownell's *D-M,* no one knows. But Griggs makes the point that his *EA* approach also neatly rationalizes subtraction of a negative number where borrowing cannot be used.

Griggs is concerned with broad transfer to subsequent mathematical topics rather than short-range transfer to a specific application or a slightly altered situation. This idea has been receiving greatly increased attention but is not new. Brownell and Moser had the same thing in mind when they pointed out that in schools where arithmetic had been taught mechanically in grades 1 and 2 pupils were unable to understand *any* explanation of subtraction. They had not learned how to use a rational or intuitive explanation. Such data

> . . . expose "readiness" for what is, namely, the ability to take on new skills or ideas, that and nothing more. . . . Their readiness is determined, not by their age or by their grade, but by the kind of arithmetic they have had. . . . By manipulating their arithmetical backgrounds we may shift the placement of any arithmetical topic over a range of several years and several grades.

Readiness is cumulative. And it is specific: a function of both topic taught and teaching method. Griggs is right to see merit in the fact that his *EA-M* method will lay the groundwork for much later work in negative numbers. But it will do so only if the later teaching builds upon the number-line argument. Current theory strongly implies the desirability of a sequential curriculum in which topics and their formulations are chosen to fit into a logically and intuitively consistent scheme, as distinct from an episodic curriculum in which bits of coordinate geometry, permutations, etc., are sprinkled through the years.

The profound observations and theories of Piaget are a necessary point of departure for any attempt to design teaching methods that conform to and accelerate child development. I shall not attempt to review his position: not only would a cursory statement be a disservice, but the sponsoring Committee has already devoted one whole conference to his work (Kessen & Kuhlman, 1962). What emerges for education is the insistence that teaching methods must link new materials to the schemata that the child already has firmly in his grasp (Hunt, 1961). Just what these schemata are and how they can be catalogued is a complex question. The assumption is that if a pupil fails on a lesson or problem he lacks certain required schemata—discriminations, concepts, motor controls, or ways of organizing and transforming information. Under our older views, the child should practice on a task just like the ultimate criterion task; if a task is difficult for him, extended practice on it is considered wasteful. According to many studies, a little practice at a later age accomplishes as much or more than a lot of early practice and with less emotional hazard. But now we have to ask whether there is some altered form of the task or some propaedeutic task that will teach needed schemata earlier. This is the rationale of Moore's instruction in reading: the child who can control his fingers and his speech can be taught to type, and this slowly generates familiarity with visual forms that the young child normally finds quite hard, and pointless, to discriminate.

This is not to say that simplifying or breaking up the task is always the key to success. Greco (1959) offers this evidence (poorly controlled): The child, aged five, is to learn that left-to-right order is reversed in a 180° rotation and conserved in a 360° rotation. The task can be simplified by presenting first many 180° trials and then many 360° trials. It is more complicated if the two types are mixed. Children learn under either training, but it is the second group that retains the learning and can cope with larger numbers of rotations. The more complex training forces the child to discriminate; in Genevan language, "to achieve a structure for the situation." The simpler situation produces only separate stimulus-response connections that lack connections to each other.

How early can we teach a concept? Consider the conservation of number, the idea that number remains the same when position or grouping changes. Piaget says that this rests on the child's attaining the idea of reversibility of the displacement, which, in turn, rests *inter alia* on his motor control and on hundreds of experiences with back-and-forth hand movements; this suggests a grand program of enriched experience in which we try to push back each stage of learning, but we know that acceleration must have some limit. Where? One view is that several radical changes of psychological state occur during development and the timing of these is biologically controlled. The most significant of these hypothesized changes after infancy are *a*) the point near age 5 where the child allegedly begins to use words to direct his own actions, so that verbal mediation for the first time makes his learning different from animal learning; *b*) the period around age 8 in which the child masters concrete operational thought, transforming information in imagination in a logically controlled way; and *c*) the early adolescent period in which formal operational thought with symbolic systems becomes efficient. The first of these hypotheses emerges from the work of Kendler (Kendler & Kendler, 1962) and Luria (Simon, 1957, pp. 115-129), the second and third from Piaget and Inhelder. Of the existence of such transitions there is no doubt, although they may be gradual rather than sharp. The question is, can they be brought about earlier?

Piaget's writings often have given the impression that he regards the timing of these developments as fixed. This occasioned great criticism from Americans who insisted that Piaget never observed a random sample of the species; the Genevans have recently been hedging their statements with references to the culture of their subjects. More than that, the laboratory has initiated a program of attempts to produce the various conceptual attainments at ages earlier than usual. These studies have had some faint success, but the results seem generally to imply that early training has little value. The best data, because of experimental ingenuity and adequate sample size, are those from Smedslund's (1961) studies of the conservation of weight and substance.

I shall report only one of his highly important observations. He taught children who had not learned that weight is invariant under deformation correctly to predict what would happen in tests such as the following: "two balls of clay are weighed and prove equal; then I roll one out into a snake; now which will be heavier, the ball or the snake?" (This is not a precise description; Smedslund's technique was better.) Success on these tests seemed to show that these children attained the conservation concept a year earlier than usual. But Smedslund made a further test. On some trials he covertly pinched off a bit of clay while making the snake. After the child predicted that the deformed ball weighed the same as the untouched ball, the ball and snake were put on the balance and the snake side went up. These trained children accepted the result. "Oh, it's lighter." "I was wrong." Control children who had attained the conservation principle in the normal way at the normal age did not. They rejected the evidence of the scale, "Did something drop on the floor?" they asked solicitously. These children trusted their internal model of the behavior of weight more than they trusted external appearances. The intensive training, however, failed to produce true conviction about conservation. There is nothing conclusive about these findings. Inhelder is presently conducting trials of various methods for training operational thought, and Smedslund also is searching for methods that will succeed more than superficially.

Mastery of a Discipline

To separate one issue from another in this discussion is like trying to count angleworms in a tin can. Readiness, preverbal understanding, meaningful presentation, transfer—all are part of the same complex. To complete this discussion I must back off and view that complex from one more angle, the purposes of the new mathematical education.

The greatest novelty in the new curricula is not the content, not the instructional methods, not the grade placement of topics. The greatest novelty is the objectives from which all else stems. I disagree violently with Martin Mayer's (1961) assertion that "in the context of the classroom nobody except an incompetent, doctrinaire teacher is ever going to worry much about the 'aims of education.'" I chide David Page for lending Mayer support by providing a know-nothing quote to the effect that "What are your objectives?" is a question on a par with "Explain the universe." Objectives had better be identifiable and explainable if one is to avoid the absurd claim that his methods achieve all ends at once, each in greater measure than any competing proposal. The new curricula, starting from the sound premise that different classroom activities reach different ends, deliberately sacrifice some ends for the sake of others.

Never before was it the aim of common-school instruction to teach mathematics. The school has usually taught arithmetic, seen as a skill of everyday commerce. The only arguments about method in arithmetic were pragmatic ones: does drill, or explanation of the number system, better guarantee ability to keep accounts or to triple a recipe? Secondary mathematics lost all sense of direction when the rug of formal discipline was pulled from under it. Part of it survived by clinging to its pragmatic justification as a prerequisite to engineering courses. The geometry course struggled unconvincingly to pose as a model for everyday "straight thinking."

If I comprehend correctly, the new curricula are trying to teach mathematics as mathematics; to the question "why?" their answer is just a shade more pragmatic than "because it's there." I have no doubt that persons in scientific and social-science pursuits need as much mathematical competence as they can get. I have serious doubts as to how far the average citizen should go in mathematics. I think it very likely that in throwing out practical topics the new programs have gone to an extreme that will produce its own form of quantitative illiteracy and provoke a counter-revolution. But as psychologist, my role is less to to say what should be done than to consider what can be done.

What does it mean to master mathematics "as a discipline?" It has taken close listening at times to be sure that the word is not a return to faculty psychology, i.e., to claims that intellectual effort is good for the mind. But the more sophisticated and constructive reformers emphasize that there are many disciplines, each of them a way of coming to grips with certain types of problems. There is obvious sense in the contention that a mathematician is more competent to solve a new mathematical problem than a bright and educated nonmathematician, and not just because the mathematician knows more theorems. He has an ability to construct models, sense connections within the model and test the internal consistency among premises and conclusions. He has a wealth of apparatus at his command—notational systems, conceptual distinctions, operations. These are used not as the computer uses a formula but as an architect uses all that has been learned from past buildings. To solve a new problem he draws from his store this and that device that might work, juggles them in the air, begins to see a coherence, discards some misfit parts and designs some replacements, and finds more or less suddenly the shape of his mathematical system. This is constructive mathematics; demonstration plays a decidedly secondary role to invention.

The greatest challenge to the psychologist from the new curricula lies here. When it is proposed to teach coordinate geometry in grade 2, the important question is whether this is a better base for some subsequent instruction than an alternative topic. When it is proposed to develop concepts of area by having children count squares within figures, we can assimilate the proposal easily. But when the objective of teaching mathematical method takes precedence over teaching mathematical results, we are led into an area where we have neither experience nor theory.

When faculty psychology was advancing the claim that mathematics equipped one as a lawyer, it was being overly general. When associationist psychology considered transfer as limited to "similar" situations, it withdrew to too conservative a position. Even the theory of transfer through generalizations is timid because it speaks of a very simple mental structure involving just one mediating sentence. The sentence is said to be "meaningful" if the pupil can coordinate it with reality. The formula for the area of a parallelogram, we have said as pragmatists, is meaningful when the pupil can determine areas of parallelograms in various positions and contexts. But the new program identifies "meaning" with the coordinations between this statement and other statements. The formula gains added meaning when the pupil sees how it is consistent with the area formulas for triangles and rectangles, and how it depends on axioms such as that shape is invariant under rotation. These propositional networks form a structure of related beliefs or insights. It is highly plausible that such structures exist in the mind, but we have not learned to appraise them and, therefore, have little insight into the process by which they develop.

One reason we know little about these matters is that we have failed to pursue our measurement of transfer far enough. As I see it, much of the aim of the new mathematics course is to develop aptitude for new mathematical learning. If so, the pertinent test is how well the student can master a new mathematical topic; for any real test we should probably use a topic that requires some weeks of study. This would be evidence of ability to go on learning on his own.

Such a measure of transfer has always been envisioned in the psychologists' definition. All the textbooks tell you that the transfer value of training A is to be tested by a design like this (where the horizontal axis represents practice time needed to reach a proficiency criterion):

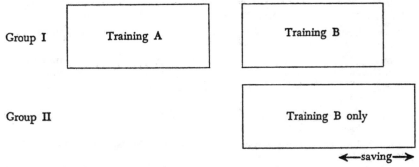

The measure of transfer is the saving in learning time. While this design is used in laboratory research, it has almost never been used in educational studies. As several of the experiments I have cited illustrate, the usual design is

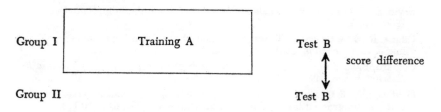

This tells us only whether we have equipped the student to make a quick adaptation to a new situation, not whether we have equipped him to work his way through a mass of material and assimilate it. Studies by the savings design will require much more effort than the usual ones. Perhaps we will find that transfer to a test and transfer to savings in new learning are highly correlated; perhaps training that achieves one achieves the other. Do you suppose, for example, that Gagné's training on how to develop a sum-of-series formula would transfer into improved ability to learn some entirely new process, such as differentiation? I doubt it. And yet that ability to come to grips with the new mathematics is certainly the product of learning, hence a function of manipulable conditions of instruction. We might well devote effort to study how one learns to learn mathematics.

The essential aim of the new curricula is to teach each subject in its true colors, as the specialist knows it: Not as a history of ancient thought, not as a collection of formulas and prescriptions, not as a display of products used by the consumer, but rather as an adventure in organizing ever-new experience. The psychologist cannot but admire the boldness of the conception and the evident art of some of the work. How to collect evidence on the effects of the programs, how to adjust his theories to conform to that evidence, and how to glean from that evidence still better educational methods: these are a challenge to *his* boldness and *his* art.

References

Ausubel, D. P. Learning by discovery: Rationale and mystique. *Bull. Nat. Assoc. of Sec. Sch. Principals,* 1961, **45,** 18-58.

Bartlett, F. C. *Thinking.* New York: Basic Books, 1958.

Bloom, B. S. *Taxonomy of educational objectives.* New York: Longmans Green, 1956.

Brownell, W. A., & Moser, H. E. Meaningful versus mechanical learning: a study in Grade III subtraction. *Duke Univ. Res. Stud. in Educ.,* 1949, No. 8.

Brozek, J. Current state of psychology in the USSR. *Ann. Rev. of Psychol.,* 1962, **13,** 515-566.

Bruner, J. S. *Contemporary approaches to cognition.* Cambridge, Mass.: Harvard Univ. Press, 1957.

Bruner, J. S. *The process of education.* Cambridge, Mass.: Harvard Univ. Press, 1960.

Bruner, J. S. The act of discovery. *Harvard Educ. Rev.,* 1961, **31,** 21-32.

Bruner, J. S., Wallach, M. A., & Galanter, E. H. The identification of recurrent regularity. *Amer. J. Psychol.,* 1959, **72,** 200-220.

Cronbach, L. J. *Educational psychology.* New York: Harcourt, Brace & Co., 1954; rev. ed.; Harcourt, Brace & World, 1963.

Ervin, S. M. Transfer effects of learning a verbal generalization. *Child Developm.,* 1960, **31,** 537-554.

Gagné, R. M., & Brown, L. T. Some factors in the programing of conceptual learning. *J. exp. Psychol.,* 1961, **62,** 313-321.

Geldard, F. A. The present status of the laws of learning. *Univ. of Virginia Rec. Extension Ser.,* 1937, **22** (No. 2), 41-45.

Goldbeck, A., & Campbell, V. N. The effects of response mode and response difficulty on programed instruction. *J. educ. Psychol.,* 1962, **53,** 110-118.

Greco, P. L'apprentissage dans une situation à structure operatoire concrete. *Études épiste. génét.,* 1959, **7,** 68-182.

Harlow, H. F. The formation of learning sets. *Psychol. Rev.,* 1949, **56,** 51-65.

Hendrix, G. Learning by discovery. *Math. Teacher,* 1961, **54,** 290-299.

Homme, L. E., & Glaser, R. Problems in programing verbal sequences. In A. Lumsdaine & R. Glaser (Eds.), *Teaching machines and programmed learning.* Washington, D. C.: Nat. Educ. Assoc., 1961, pp. 486-496.

Hunt, J. McV. *Intelligence and experience.* New York: Ronald Press, 1961, pp. 266 ff.

Judd, C. H. *Education as cultivation of the higher mental processes.* New York: Macmillan Co., 1936.

Katona, G. *Organizing and memorizing.* New York: Columbia Univ. Press, 1940.

Kendler, H. H., & Kendler, S. Vertical and horizontal processes in problem solving. *Psychol. Rev.,* 1962, **69,** 1-16.

Kessen, W., & Kulhman, C. Thought in the young child. *Monogr. Soc. Res. Child Developm.,* 1962, **27,** No. 2.

Kieslar, E. R., & McNeil, J. D. A comparison of two response modes in an auto-instructional program with children in the primary grades. *J. educ. Psychol.,* 1962, **53,** 127-131.

Krumboltz, J. D., & Weisman, R. G. The effect of overt versus covert responding to programed instruction on immediate and delayed retention. *J. educ. Psychol.,* 1962, **53,** 89-92.

Mayer, M. *The Schools.* New York: Harper & Bros., 1961.

Miller, N. E. *Graphic communication and the crisis in education.* Washington, D. C.: Nat. Educ. Assoc., 1957.

Miller, N. E., & Dollard, J. *Social learning and imitation.* New Haven, Conn.: Yale Univ. Press, 1941.

Osgood, C. E. *Method and theory in experimental psychology.* London: Oxford Univ. Press, 1953.

Silverman, R. E., & Alter, M. Note on the response in teaching machine programs. *Psychol. Rep.,* 1960, **7,** 496.

Simon, B. (Ed.) *Psychology in the Soviet Union.* Stanford, Calif.: Stanford Univ. Press, 1957.

Skinner, B. F. The science of learning and the art of teaching. *Harvard Educ. Rev.,* 1954, **25,** 86-97.

Smedslund, J. The acquisition of conservation of substance and weight in children. I-VI. *Scand. J. Psychol.,* 1961, **2,** 11-20, 71-87, 153-160, 203-210.

Spence, K. W. The relation of learning theory to the technology of education. *Harvard Educ. Rec.,* 1959, **29,** 84-95.

Exercises: Issues Current in Educational Psychology

1. Cronbach contends that the greatest novelty in the new curricula is "the objectives from which all else stems." Discuss the validity of this contention citing examples to illustrate your position. How would changing objectives affect your search for psychological bases for teaching decisions?

2. Differentiate between "guided discovery," "discovery," and "rule-and-example" teaching. What psychological principles would tend to support which type of teaching? Would you expect one pattern of teaching to be more effective for specific types of learning? Provide specific examples from mathematics to illustrate your response.

3. Why do psychological principles seem to "come and go"? Is this due to changes in the nature of learners (or learning), to the nature of what is to be learned, or to something independent of both?

4. Cronbach lists seven "principles of learning" accepted by educational psychologists, circa 1940. Find a mathematics textbook of that period. Can you identify pages that seem to be applications of each of these seven principles? Which of these pages might still appear in a modern mathematics textbook? Which could not?

5. Which, if any, of these principles do you feel would not be supported by educational psychologists, vintage 1970? Generate a comparable list of principles you feel would be accepted by educational psychologists, vintage 1970.

6. What assumptions do the proponents of discovery teaching make about the way children learn?

7. What is discovery learning? What is Cronbach's "stand" on discovery? Choose a topic from mathematics and illustrate what you believe to be Cronbach's position concerning the role of discovery.

8. Give two alternative ways to teach a mathematics topic. Which way best deals with cumulative readiness for other topics in mathematics as Griggs argues is an advantage for the *EA-M* method? Illustrate.

9. What do you see as the reasons for teaching mathematics? Do your reasons justify the content now being taught or the testing used to evaluate mathematics instruction? Explain.

10. What is transfer? Is transfer a significant factor in the learning of mathematics? Give an example of transfer for a

mathematics topic and describe a procedure for evaluating a student's ability to transfer the topic. Give test items.

11. Several states now require evidence of students' mastery of mathematics in order for a school to qualify for state monies for education. Discuss what type(s) of evidence would be appropriate to define mastery for this purpose.

Unit 2

The Nature
of Mathematics Learners

Individualization is a key word in educational discourse today. Held as an humanistic ideal, individualizing instruction assumes an aura of romanticism when considered as the answer to many pressing educational problems. In point of fact, little is known which the teacher can use to prescribe instructional strategies on an individual basis. As a result, much of what is done in the name of individualization is closer to isolation—letting the student work alone at his own pace. This is a particularly frustrating state of affairs in an era in which computer technology holds the promise of successfully treating the classroom management problems associated with individualized instruction. Until individual learning styles and the associated symptoms are identified, the teacher's domain for decision making appears to be restricted to controlling the rate of presentation. Even intuitive judgments of what will serve as a motivating context for learners are seldom made with the individual in mind.

Several characteristics of the mathematics learner are useful to the teacher in his decision making. Brownell comments on the progressive, incremental nature of developing understanding in mathematics. His analysis of the relation of process and product learning establishes a basis for using drill to strengthen computational skills. The discussion of connectionism, dating from 1944, supplements Cronbach's historical analysis of the preceding section.

Davis's analysis of Jennifer's learning difficulties reinforces and extends Brownell's commentary on appropriate strategies for the learner who has been unsuccessful in earlier instruction. How can you sharpen your skills as a diagnostician of learning difficulties? Without knowing the root of the difficulty, how can you successfully prescribe treatment?

The final segments are devoted to the developmental psychology of Jean Piaget. Piaget's work has extended over several decades and has been the subject of extensive replication by researchers here and abroad. Piaget's conception of intellectual growth and development reveals his background orientation of biological science. Three scientific concepts basic to Piaget's model are assimilation, accommodation, and dynamic equilibrium. The series of excerpts from Piagetian interviews provide specific examples of the evidence Piaget used to build his theory. These are also good examples of the questioning skills needed for the diagnosis demanded by the two preceeding articles. They should be read in close conjunction with the following article of Adler's.

The final article by Adler outlines major characteristics of Piaget's empirically supported model and establishes some major cautions for the interpretation and classroom application of Piagetian ideas. Growth and development psychologists have a tendency toward recommending the teacher passively accept the maturation of the child as a limitation on what should be taught. Readiness to learn will come as the child grows intellectually. Does Piagetian psychology support this passivity? What does readiness mean in Piagetian terms? How do you decide what are appropriate learning and testing experiences for children who have not attained the final maturity level delineated by Piaget? These questions are evidence of some of the real payoff of familiarity with Piagetian psychology—it suggests many fascinating avenues of exploration.

The Progressive
Nature of Learning
in Mathematics

William A. Brownell

There is a close relation between our teaching procedures and our conception of the learning process. As the first step in teaching we more or less carefully determine our objectives. Then, to help our pupils attain these objectives, we select our explanations, our types of practice materials, and our applications largely in the light of our theory of learning. It follows therefore that his view of learning is a critical part of every teacher's professional equipment.

Learning as Connection-Forming

The conception of learning which has prevailed in American education for more than a quarter-century—less so now than formerly—is part of the psychological system known as connectionism. According to this view, all learning consists in the addition, the elimination, and the organization of connections—this, and nothing else. These connections are formed, or broken, reorganized, between situations and responses. The process of teaching, then, comprises the following steps:

1. Identify for the learner the stimuli (or the situation) to which he is to react,
2. Identify the reaction (or response) which he is to make,
3. Have the learner make this response to the situation under conditions which reward success and which punish failure,
4. Repeat step 3) until the connection has been firmly established.

The connectionistic view of learning has not, in my opinion, been very helpful to teachers.[1] Advocates of connectionism could hardly accept my

William A. Brownell, "The Progressive Nature of Learning in Mathematics," *The Mathematics Teacher,* Vol. XXXVII, No. 4, 1944, pp. 147-157.

[1] I have never been able to find much that is wrong (demonstrably unsound) in connectionism. Most of the direct attacks, theoretical and experimental, seem to me to be rather futile: they have failed to show that learning is anything other than the formation of connections (if the term "connection" be interpreted as broadly as connectionists interpret it). As a matter of fact, I suspect that, neurologically at least, something very much like the processes suggested by connectionists actually occurs in learning (this, in spite of the research of Lashley and others). Objection here is raised to connectionism purely on the ground, as stated above, that it has not been helpful in the practical business of educating children. Perhaps an analogy is in order. The layman reads that all matter is reducible, according to modern physics, to electrons, neutrons, etc., in a word, to nonmatter. He can accept this view of matter as a fact. At the same time, acceptance of this view does not in any way affect the manner in which he deals with matter, however essential this view is to the work of the research physicist.

evaluation. If they were to concede that their view of learning has not always produced the best results, they would insist that the deficiency lies, not in the theory, but in the user. They would say, as many of them have said, that the theory is sound and adequate, but that it has been misinterpreted and misapplied.[2]

In this paper I want to consider with you four weaknesses in classroom instruction in mathematics. (They are by no means confined to the teaching of mathematics.) These weaknesses persist after thirty years and more of connectionism. Whether they are still with us because of the connectionistic view of learning or in spite of it, it is impossible to say. It is however possible to show how this view of learning seems to support, even to demand, the malpractices which I shall discuss. It is for this reason that the connectionistic view of learning will come in for unfavorable comment.

I am well aware that I am talking, not to professional psychologists, but to teachers of mathematics. Indeed, it is precisely because I *am* talking to teachers of mathematics that I speak as I shall. Perhaps no other subject in the curriculum so much as mathematics has suffered from the general application or the general misapplication of connectionistic theory. But my remarks will not all be negative. On the contrary, as the subject of the paper implies, I shall try to substitute positive notions and shall try to sketch a different view of learning which may be more useful to teachers.

Four Instructional Weaknesses in Mathematics

The four instructional weaknesses to which connectionistic theory has contributed directly or indirectly are:

1. Our attention as teachers is directed away from the processes by which children learn, while we are over-concerned about the product of learning,
2. Our pace of instruction is too rapid, while we fail to give learners the aid they need to forestall or surmount difficulty,
3. We provide the wrong kinds of practice to promote sound learning,
4. Our evaluation of error and our treatment of error are superficial.[3]

1. Process vs. product.—I have said that connectionistic theory leads us to neglect the processes by which children learn. This is so because, unlike the *product* of learning, the *process* of learning seems scarcely worthy of attention. The reasoning is somewhat like this: all learning is

---

[2]Of all the exponents of connectionism Gates has argued most cogently in this vein. Arthur I. Gates, "Connectionism: Present Concepts and Interpretations." *The Psychology of Learning,* Chapter IV. Forty-first Yearbook of the National Society for the Study of Education, Part II. Bloomington, Ill.: The Public School Publishing Co., 1912.

[3]In the literature of connectionism I can find statements which contradict all these charges. That is to say, connectionists recognize, in theory, the evils which are listed above. But the habit of thinking of learning as connection-forming almost inevitably oversimplifies learning and affects teaching adversely. Perhaps no better example of this tendency is to be found than in the books *The Psychology of Arithmetic* and *The Psychology of Algebra,* both written by the originator of connectionism, Professor E. L. Thorndike.

the formation of connections; or, the process of learning *is* the making of connections. Connections all being basically alike, the process is the same for all learning. Hence, we need only to make sure that the right connections are established; they will then necessarily lead to correct responses. We know what the correct responses are; we identify them for learners, along with the appropriate stimuli, and we provide practice until the desired connections between the two are formed. The process of learning, which is to say, connection-forming, takes care of itself. From all this we come to teach by telling children or showing children what to do and then by seeing that they do it.

We tell children that 2 and 5 make 7; we show them how to divide one fraction by another; we give them the rules governing signs in algebraic operations; we furnish them the facts of the Pythagorean Theorem. Practice follows to establish the connections. When our pupils demonstrate that they have the desired connections by producing the correct responses, the teaching job is done.

Now, more is the pity, some children try to learn mathematics according to this simple pattern. I shall have more to say about these children at a later point. It is pertinent here, however, to note the attitudes which such children develop toward mathematics. The correct answer is their sole consideration. Let them, by no matter what curious manipulation of symbols, arrive at an answer which agrees with that of teacher or textbook, and they feel that they have met all requirements of the situation. Change in the slightest degree the conditions in which the mathematics occurs, and they are helpless. Challenge an answer even when it is correct, and they have no way to prove it. To tell them that mathematics, whether it be arithmetic or algebra or geometry, is a system of logical relations is to speak in a foreign language.

But a large per cent of children do not learn mathematics in this simple, blind way, even when the teaching might seem to encourage such learning. Instead, they try somehow to put sense into what they learn. They may be told—and told time and again—that 2 and 5 make 7; but they forget it. When the forgetting becomes too embarrassing, they try something beside memorization. They turn 2 and 5 around to make 5 and 2, which for some reason they may know better; or they count one number onto the other as a base; or they see 2 and 5 as the same as 3 and 4, with a sum of 7.

You are probably all familiar with the girl reported by Stephenson, a girl who has many mathematical relatives in every community in which I have lived. This girl found verbal problems too much for her; or, they were too much for her until she devised some simple rules: when the problem contains several numbers, you add; when it contains two long numbers, you subtract; when the larger of two numbers exactly contains the smaller, you divide; otherwise, you multiply. This girl, like her many previously mentioned relatives, has been held up as a conspicuous example of stupidity. I cannot agree. I believe, instead, that she showed an extraordinary degree of originality and resourcefulness. As a matter of fact, barring computational errors, she would, by her procedures, get the correct answers for the majority of problems in texts for the lower and intermediate grades. And if she got the wrong answers? Well, so did the other children who had different "rules." It just happens that the processes used by this girl are not mathematical, and she was supposed to be working in the area of mathematics. But mathematics as such meant nothing to her; she had not learned the tricks of the trade; so, she invented some of her own.

What I have been doing, you will have recognized, is to illustrate negatively the importance of *process* in learning. Let us now approach the matter positively. Consider the example 42 + 27. There is not just one way to find the answer, as we sometimes naively assume. There are almost numberless ways:

a. One may count out 42 separate objects, lay them aside, count out 27 more similar objects, lay them aside in a separate group; and then find the total by counting all the objects by 1's, starting with 1.
b. One may use objects as in a, getting two groups of 42 and 27, and again count by 1's, but start with 43 or 28 instead of with 1.
c and d. One may use objects as in a and b, but count by larger units than 1's, as by 2's or 5's.
e. Still using objects, one may set up 42 as four groups of 10 objects each, with 2 over, and 27 as two groups of 10, with 7 over; he may then count the tens, getting 6, and the ones, getting 9, to yield the total of 69.
f to j. One may use any of the first five methods, substituting marks or pictorial symbols for actual objects.
k. One may count the 27 onto the 42 abstractly, or the 42 onto the 27.
l and m. One may copy the abstract numbers and get subtotals in the two columns by counting abstractly or with marks.
n. One may know the separate combinations and add directly: 2 and 7 are 9; 4 and 2 are 6; total, 69, without knowing anything about the composition of the numbers dealt with—a purely mechanical stunt.
o. One may proceed as in n, but be fully aware of the nature of numbers and of the process of addition.
p to N. One may do any one of the many things which this audience would report as their processes, such as: 1) direct and immediate apprehension of the total; 2) adding 20 to 42, and then adding the remaining 7; 3) adding 40 and 20, adding 2 and 7, and then combining; 4) adding from the left, with a preliminary glance to make sure that no carrying is involved; and so on, and so on.

All these procedures, and others not here catalogued, are entirely legitimate. All of them, except the last few, may be found in actual use in classrooms in which the procedure for adding such numbers as 42 and 27 is being taught. The teacher knows the *product* she is seeking to attain, namely, skill in adding two-place numbers without carrying. If she thinks of learning purely as the formation of connections, she is apt to oversimplify the learning task. She will tend to show her pupils how to add the digits in the two columns and where to place the partial sums; and then she will rely on practice to establish the needed connections.

I have tried to show that identification of the learning process with the formation of connections, however valid for ultimate psychological and neurological theory, is not useful to teachers. Teaching is the guidance of learning. We can guide learning most effectively when we know what the learners assigned to us really do in the face of their learning tasks. In a word, we as teachers can be helpful in guidance to the degree to which we know our pupils' processes. I do not mean that the product of those processes is no concern of ours; but I do mean that processes are of at least equal importance with products. The teacher who knows the product which is to be finally achieved, but who also knows how to discover, evaluate, and direct the processes of her pupils as they approach this goal—that teacher is probably a good teacher. Moreover, thinking of learning as the formation of connections would not make her a better teacher.

2. Over-rapid instruction.—So much for the first objection to the connectionistic view of learning: it takes us as teachers away from our main stock in trade, namely, the processes by which children learn. The first objection is closely related to the second: it tends to make us hurry unduly the pace of instruction and it discourages us from supplying to children temporary aids and procedures which they need for sound learning. In a word, we are led to think that children can complete their learning at a single jump.

Without reciting them again, let me recall the list of processes by which one may find the sum for 42 and 27. These processes were described in the earlier place only to establish the existence of various possible processes. Let me cite them again here for another purpose. The processes were arranged by plan, in a roughly ascending order of complexity, maturity, and abstractness. It is surely obvious that the child who counts by 1's 42 objects, then 27, and finally the total of 69 is acting more simply and concretely and less maturely than does the adult who apprehends at a glance the sum of the two abstract numbers. And the other processes listed can be posted at intermediate points in the scale of maturity and abstractness. Any given child may be at any point in this scale. Furthermore, his degree of mastery of his process, whatever it is, may vary from inexpertness to striking proficiency. Indeed, a child whose procedure (counting by 1's, for example) is very low in the scale of maturity may out-score in rate and accuracy another child whose procedure is at a higher point in the scale. Failure to note this fact is one of the penalties we pay for neglecting process in favor of product only, and it constitutes a large source of error in the evaluation of learning.

We have come to accept the typical curves of learning as picturing all that goes on in learning. The accuracy curve, for example, mounts rapidly at first and then slower and slower. The implication is that the child is getting the desired connection established just about as is portrayed in the curve. But when we examine into the behavior of the child, we find the situation to be much more complex than this. He may practice a given procedure for a time and then desert it for another, and this for another, and that for still another. A more valid picture of his learning, this time plotted in terms of process, would look something like a series of steps, each successive one somewhat higher in the scale of maturity than the preceding one. In a word, the learner progresses by traversing a series of stages in thinking. Each stage serves its purpose for a time, but is superseded by a more advanced stage. As each stage is abandoned for the next, the earlier stage is not forgotten or gone—its pattern is not eradicated from the nervous system. Instead, the older procedure is overlaid by another, and the old neural pattern remains for use if for any reason the more recently acquired procedure does not function smoothly.

This description of learning, while it may be consistent with connectionism, is certainly not suggested by connectionism. On the contrary, the connectionistic view of learning leads us to give the child at the outset the form of response which we want him ultimately to have. And we are inclined to do this quite without regard to his attained stage of thinking when we present the new learning task. At best, the result is pseudo-learning, memorization, and superficial, empty verbalization.

Let me illustrate what I mean. I once asked a third-grade teacher to send me for interviews the three poorest arithmetic pupils in her class. George was one of the three. He was described as an almost certain failure. After some preliminary exercises, I put before George the three digits 8, 7, and 9 arranged for column addition, and asked him to find the sum. Listen to him: "8 and 7 are 15, and 9 are . . . I don't know." Then, in the reverse direction: "9 and 7 are 16, and 8 are . . . I don't know." After a pause George looked at me and asked, "May I count?" I told George he might, and promptly he mumbled, "8 and 7 and 15, 16, 17, . . . 24." When I asked him to add upward, he produced the correct answer with equal quickness by counting the 8 to the sum of 9 and 7. Obviously, George did not understand bridging of the decades, save as this could be done by counting.

But why did George ask if he might count? Clearly because his teacher had told him that he must not count, with what penalties I do not know. George obeyed her, and the reward for his obedience was failure. Parenthetically may I say that one of the three *best* pupils sent from the class for interviews counted not only to bridge, but counted to find the sum of each pair of digits. This last child, singled out for praise, was really at a lower stage of thinking than was George; but the teacher did not know it. She had asked these children to perform at a level or stage of thinking of which they were then incapable, and they had reacted according to their natures.

To return to George and to repeat what I have already said— this lad was at a very immature level with respect to bridging: he had to count if he was to bridge at all. I said to George, "George, how many are 8 and 7?" He said at once, "15." I asked, "How many more does it take to make 20?" He said promptly, "5"; and then after an instant his face lit up and he said, "And 4 more are 24!" Then, "Will it work like that every time?" I asked him to add the same numbers upward, and he said, "9 and 7 are 16; . . . and 4 are 20, and 4 are 24! Can I do that with all of them?" I wrote several examples which called for bridging 20, and he solved them quickly, understandingly, and triumphantly by his new method.

What had I done? Well, I had helped George to the next higher stage in thinking, a stage for which he was ready, but a stage which he had not discovered for himself and which he was unlikely to find, so long as instruction consisted only in telling what he must not do.

Should George have been left at the stage to which I had helped him, always to bridge 20 or some higher decade by splitting a number? No; he should have been led next to understand the principle of adding by endings, and still later to think immediately and directly of the total of the last partial sum and the last digit. Sound learning required that he traverse these intermediate steps; he was getting nowhere by his own devices. Unfortunately circumstances prevented my helping him discover the next steps, and I fear that his teacher may not have provided the needed assistance either.

I have spent all this time on George because his case illustrates so clearly the dangers of too rapid a pace of instruction. It illustrates too the need for temporary or intermediate processes, methods, and devices which are now commonly kept from children to their detriment. These temporary aids have a bad reputation in education. I am sure you have

often heard the dictum: "Never form a habit which must later be broken."
The addition of the qualifying phrase "Other things being equal,"
whatever its intent, does little to soften the ban against their use. Indeed,
so fully are school officials and teachers persuaded of the evils of these
aids that they are to be found in few textbooks, and there in insufficient
amount. I do not mean however to imply that these aids are totally
unknown to the classroom. Good teachers use them—sometimes openly,
perhaps more often when the supervisor or principal is not likely to
appear.

So long as we think of learning as a simple, straight-line development,
the warning against temporary aids makes sense. These aids seem to
contribute little; they increase the number of things to be learned; they
tend to be retained after they have long outlived their usefulness (if any).
But we must cease to think of learning, and certainly of learning in
mathematics, in this manner. We do not seek merely to develop a few
mechanical skills always to be used precisely as they were learned. The
purpose of mathematics, whether in the elementary school or in the high
school, goes far beyond the establishment of mechanical skills. The
ideas, principles, generalizations, and relationships which are taught, *as
well as the skills,* are intended for purposes outside themselves and for
use in situations quite unlike those in which they are learned. We teach
quantitative mathematics, for example, as a system of thinking by which
to manage and control number and quantity, not alone as presented in
textbook problems or as presented in the classroom, but however and
wherever and whenever presented. In a word, we strive to teach
understandings. When the goal of understanding is accepted, the
function of temporary aids is seen in its correct perspective. Such aids
contribute meanings when meanings are needed; and the more
meanings, the deeper the understanding, and the greater the chances of
successful transfer to new and unfamiliar situations.

In making a case for temporary aids I may have failed to indicate all
that should be included in this category. My illustrations have all been
what may be called "lower-order thought procedures." But there are
others as well. Here belong also the techniques and materials commonly
embraced by the term "sensory aids"—drawings, pictures, maps, films,
diagrams, slides, solid articles intended primarily for classroom use, the
geometric forms of architecture, and so on.

The argument for the abundant use of such aids is precisely the same as
that for lower-order thought procedures. Learning is progressive in
character. The abstractions of mathematics are not to be attained all at
once, by some coordinated effort of mind and will. Instead, we must start
with the child wherever he is, at the foot of the ladder, or at some point
higher up. Well chosen sensory aids reveal the nature of the final
abstractions in a way which makes sense to the child. If he can work out
the new relationships in a concrete way and can himself test their
validity in an objective setting, he has faith and confidence at the start;
and he is the readier to learn with understanding the more abstract
representations of mathematics. Sensory aids, like many so-called
crutches, are then not only admissible under the conception of learning
which I am outlining: they are obligatory.

On the other hand, the connectionistic view of learning does not
predispose the teacher to employ temporary aids to the extent to which he
should employ them. Instead, for reasons which I have already

mentioned, it leads him too quickly to abstract practice or drill. And this is the third objection I have listed for consideration.

3. Faulty practice.—It would be false to accuse connectionists of spreading the gospel that "Practice makes perfect."[4] Nevertheless, the teacher who accepts their view of learning comes easily to rely upon drill as his exclusive or major teaching procedure. Listen again to the litany: Identify for the learner the situation to which he is to react, and the response he is to make to the situation; have him make the connection under satisfying conditions; have him exercise the connection until it is firmly established. Does not this *sound* like an exhortation to drill?

The apparently innocuous statement, "Practice makes perfect," is full of dynamite because it conceals important issues. Practice *does* make perfect in one sense of the words "practice" and "perfect"; but it makes for superficial learning in another sense of these same words. I can make my point by using a crude illustration.

Suppose that I, an amateur at golf, dub my drive: the ball barely trickles off the tee. What do I do? Practice? If so, in what sense? Well, if I practice in the sense of *repeating* my act, I grasp the club the same way, stand the same way, and swing the same way. With what result? I shall certainly get the same result—with this difference: I shall have become a bit more proficient in my poor drive. That is to say, there will be a little more economy and ease in my movements; I shall be more certain of what will happen; I may swing a little more quickly and precisely. Now, suppose I continue my repetitive practice over a period of time. Eventually, I shall become the most efficient poor golfer on the course, for repetitive practice will have brought its consequences.

But, you say, no golfer would do anything so silly as this; he would know what would happen. At first his practice would not be repetitive, but varied. That is to say, he would try to *avoid* the first combination of movements which produced the poor drive. He would stand differently, hold the club differently, swing differently, and so on. Only when he had arrived at the best possible combination of movements, as judged by the kind of drive it produced, would he start repetitive practice. And he would repeat purposely, because he would know that repetition could now give him the efficiency he desired.

But you are teachers of mathematics. What does this little excursion into golf have to do with the teaching of mathematics? At the risk of being dogmatic, I should say a good deal. As a matter of fact, we learn motor skills (the drive in golf) about as we learn abstractions (mathematics). It is only because we cannot directly observe the activities in ideational learning that we come to think them different for motor learning. However, the appearance of difference vanishes when we adopt appropriate techniques for observation. Then we see the essential sameness in the case of motor skills and of abstractions. In both instances learning is characterized by the organization of behavior at successively higher levels. It is because of this essential sameness that the illustration from golf may prove helpful in showing the real function of practice in learning mathematics.

We are now ready to formulate two statements which should supplant the familiar "Practice makes perfect." a) *"Varied practice leads to the*

---

[4]Some connectionistic accounts of learning contain explicit warnings against over-reliance upon repetitive practice.

discovery of the right combination of movements and ideas." b) *"Repetitive* practice produces efficiency, but at whatever level of performance the learner has attained." The proviso in the second statement is highly important: if repetitive practice is introduced prematurely, the learner is "frozen" at his level of performance. He steadily becomes more proficient at an undesirably low level of maturity. Repetitive practice cannot move the learner to a higher or more mature level. If, under conditions of drill or repetitive practice, the learner does actually move on to a higher level, the credit does not belong to the drill to which he has been subjected. An examination into his behavior will reveal that he has deserted the prescribed repetitive practice and has struck out for himself into varied practice.

It follows from what I have just said that in the end we defeat our own purposes by introducing repetitive practice too soon. In the early stages of mathematical learning we need to institute activities which will enable the learner to explore understandingly the new area which he is entering. The learner is not exploring this area when he does nothing but repeat what he has been told or solve problems in the way in which he has been shown. Continued practice of this kind can yield nothing more than superficial learning: his efficiency may mislead us into thinking that he has a more thorough grasp than he actually has.

It also follows from what I have said that there is plenty of room and a great deal of need for repetitive practice in mathematics. Extremists in education who have reacted quite properly against premature drill do not correct the evil when they say that drill has no place at all in modern teaching technique. Such individuals would certainly engage in repetitive practice in golf; they would not for a minute believe that the limited insight of one lucky drive would give them command of the stroke. Yet, they would abolish drill in ideational learning. But drill is an inappropriate teaching procedure only when it is called upon to do what it cannot do. It is entirely appropriate when the goal is efficiency.

4. Remedial teaching.—We come now to the fourth objection to the prevalent conception or misconception of learning: uncritical acceptance of connectionism is likely to lead to the misinterpretation of error and to the use of inappropriate remedial measures.

When a child makes a mistake, it is easy, but misleading, to say that he has made the wrong connection or that he has failed to make the correct one. And it is easy, but wrong, to assume that that mistake can be remedied merely by showing the child the correct response and having him practice until a connection is formed. This analysis of error and the proposed remedy might suffice if the learner were dealing with nonsense syllables or with puzzles which he could solve only by accident.[5] If the

---

[5]As is well known, the connectionist *as a connectionist* (that is, when he is concerned primarily with theory) has experimented chiefly with just such experimental problems. Even in such learning situations as he sets for the learner the connectionist would, if he but attended to what the learner does (his process), discover that learning is much more complicated than it appears in his account in terms of rate and accuracy. He would discover, even in the case of nonsense syllables and of mazes, that the learner approaches the final goal of efficiency through a process of reorganization at successively higher levels. Gates has recognized this fact to a far greater extent than have most connectionists. See: Arthur I. Gates, *Psychology for Students of Education*, Revised Edition. New York: The Macmillan Co., 1930. Chapter XI.

learner fails to recall one of the syllables or if he makes the wrong turn in a maze, we give him the correct syllable, or we tell him his error in the maze or show him the correct turning. Remedial instruction in such learning assignments seems to be simple indeed: tell or show, and practice.

But the classroom should present exceedingly few learning tasks of these kinds, particularly in mathematics. Of course the learner must memorize arbitrarily predetermined characters and symbols, and in this case remedial instruction consists primarily in telling or showing and then in practicing. But, typically, learning tasks in mathematics are far more complex since they involve meanings and understandings.[6] I believe you will agree that most errors in mathematics are the result, not of imperfectly learned symbols, but of incomplete understandings, of inappropriate thought processes, and of faulty procedures.

Consistently throughout this paper I have stressed the progressive character of learning: the learner moves from level to level in thought processes, each successive level being more mature, more abstract, more adult-like than the preceding. Except in the case of imperfect mastery—and drill is then the remedial measure—except in such cases errors come from failure to traverse the stages and levels of thinking in an orderly fashion. Called upon to perform at a level higher than any he has yet attained and given no guidance to reach the higher level, the child has but three courses of action open to him. a) He can refuse to learn. Refusal may take several different forms. One form is, "I won't." Under ordinary conditions of schooling this form is not common. Another form is, "I can't"; a third is, "I don't want to," or "I don't care." The result of refusal, by whatever form, is indifference toward mathematics or dislike of it, which may be accompanied by widespread feelings of frustration.

If the child does not refuse to learn, he may adopt a second method of extricating himself from his predicament. b) He may do his best to perform as he is asked to perform. This method is probably the commonest, and it reflects the effects of years of training in docility. Of course the child cannot actually do what he is supposed to do, but by blindly following rules and by profiting from model solutions he may *seem* for a time to be successful. His answers are correct, and he gets them with reasonable promptness. This evidence of learning, based upon the criteria of rate and accuracy, is spurious. The skill which is developed will not be useful except in the situations in which it is learned, and after a short while even this degree of skill is gone.

c) The child's third method of surmounting his difficulty is to fool his teacher, to continue actually to perform at his attained (but not at the expected) level, but to conceal this fact. This method is successful when the child develops enough proficiency with his lower-order procedure to equal the performance of other children who are higher in the scale of maturity. And it is not impossible to attain this degree of proficiency. Many children do, and the fact is not discovered until later, perhaps a year or two, when with the growing complexity of learning tasks the low-order procedures are no longer effective.

---

[6]The connectionist of course agrees, but he has what to him is an explanation of the complexity. The learner is merely forming many connections at the same time, concomitantly and in succession. I can agree that this may possibly be so. At the same time I find little in this view which provides guidance to the teacher.

It should be clear that no one remedial technique will be successful with these different types of disability. The child whose attitude toward mathematics has been ruined needs to have that attitude corrected. The working of masses of unenlightened and unenlightening examples and problems will not reach the source of difficulty. If the undesirable attitude arose because of inability to understand and of a consequent series of failures, the child's attitude will improve only when he understands and when he has had ample experiences of a successful kind.

Similar treatment may be necessary for the child who has developed spurious skills, only to lose them. More drill will serve only to revive for a time the skills which never were worth the trouble to cultivate. It is for this reason that so-called refresher courses in mathematics in the high school must frequently fail of their object. What most high-school students need is not more of the kind of instruction which produced their deficiencies, but a type of instruction of which they have not had enough. If the giving of this instruction means that the usual systematic mathematics courses in the high school must be sacrificed for many children, what of it? With the appallingly weak foundations many high-school students have in earlier phases of mathematics, they can hardly profit from the more advanced phases to which they have been traditionally exposed.

I have considerable sympathy for the child who has adopted the third way out of difficulty, that of remaining at a low level of performance and of developing proficiency therewith. After all, whatever he does, and however far he may be from the desired level of performance, he knows what he is about. In this connection let me recall the case of George who had trouble with bridging the decades in addition. The way to secure a higher level of thought process in such cases is certainly not to deny the child knowledge about that higher process, at the same time forbidding him the only process he has. Nor is it any more effective to assign large bodies of drill, for, in the case of less honorable children than George, the drill examples are but so many more invitations to increase facility of performance at a low level. Here as elsewhere, remedial measures must accord with some particular kind of deficiency.

I cannot quit this matter of remedial instruction without commenting upon certain practices which I have witnessed. I am both amused and irritated by the behavior of some college professors of mathematics when their weaker students come to them for help. Observe them, or observe a typical pair. The professor greets the confused student pleasantly; he sits down with him at a table, turns to the section of the text which is causing trouble, and neatly copies an example on a nice, clean sheet of slick white paper. Then they go to work. But before we come to the "work," let us note that the professor copies the example with a *pen,* and he does all his work with a *pen.* Now perhaps memories of my own difficulties in college mathematics make me hypersensitive, but I resent that pen! I claim that the pen adds insult to injury. Consider the gulf its use establishes between student and professor. The student does his mathematical work with a pencil which is equipped with a large, competent eraser—and he uses that eraser frequently and vigorously. But the professor! he is so sure of himself that he can use a pen, thereby making a record which cannot easily be altered—only of course the professor knows that he won't have to alter his record. So does the student, and this knowledge may shake still more his already wavering self-confidence.

But I have admitted that my prejudice against the pen may be purely personal. So, let us get back to the work. What happens? The professor writes out each step in the solution calmly and certainly, meanwhile accompanying himself with a monologue. I cannot say that he is carrying on a conversation because the student contributes no verbal comment. Indeed, it is to be questioned whether he contributes anything at all, including understanding of what is going on. At the conclusion of the exhibition the professor settles back well pleased with himself, and says mildly "Well, there it is. Do you see how to do it?" Courtesy alone would require the student to say that he did see it, even if the stupefaction which magic produces did not render him incapable of more than nodding his head weakly or saying merely, "Yes."

Whatever name we give to this seance, we cannot call it remedial instruction. It might be a good idea to deny the professor all use of pencil and paper—and certainly of pen and paper! The student has come for *help,* not for a demonstration of the professor's skill in mathematics. If the professor were unable to write out his own processes, he might give the student the kind of help he has come for. The student, not the professor, should do the work. He should go as far as he can on his own; when he can go no further, he should be questioned and guided through questions to locate his difficulty and to analyze its nature. Through continued questioning he should be led to suggest possible next steps and then to evaluate these steps himself. But at all stages the student should be required to make use of his own knowledge (to the extent that he has any) and he should be allowed to identify his deficiencies himself and to feel that he is making progress by his own efforts. Remedial instruction of this kind is worthy of the name, and the results justify the time and energy that must be expended to secure them.

So much for the four instructional weaknesses which I listed for discussion, weaknesses which, if they cannot be attributed to the connectionistic view of learning, have certainly not been dispelled by the general acceptance of this view in American education. At the outset I promised that my comments would be constructive as well as critical. I think I have kept my promise. For a conception of learning which may be helpful so long as we deal with the most uncomplicated types of learning I have offered a substitute which may be more helpful for the kinds of learning which are involved in mathematics.

This latter conception stresses the notion of progressive reorganization. It emphasizes the essential continuity of learning. It points out that the learning of relationships and the development of meanings take time which is filled with suitable activities. It defines the kinds of practice which are appropriate at different stages in learning. It guarantees that children will not be hurried toward empty verbalizations but will be directed toward useful abstractions. In a word, it is a conception which provides insights into the course of learning which children must pursue if they are to attain the approved objectives of mathematics. I commend this conception to you teachers who are charged with the responsibility of guiding children to a meaningful and intelligent grasp of mathematics.

**Exercises: The Progressive Nature
of Learning in Mathematics**

1. In your view, what mathematics in the classroom should be learned by following the four steps given by Brownell for connection-forming? Why? Illustrate your position with examples. What possible weaknesses do you see for your position? Illustrate.

2. The Brownell paper was read in 1944. What is the relationship between connection-forming and the eight learning types of Gagné?

3. What is your reaction to the example of the girl reported by Stephenson who had developed a procedure for word problems? Can you find a similar example in your own learning of mathematics? Or your students'? Illustrate.

4. Brownell illustrates a multitude of ways to do the problem 42 + 27. Do the same thing with some mathematics you expect to teach. (If you have never talked with children regarding the example you have picked you will be at a disadvantage. However, try to come up with at least ten distinct, believable ways.)

5. Brownell criticizes some mathematics teaching for being "over-rapid instruction"; yet Gagné contends that once the prerequisite knowledge is known by the student, the learning of a principle is nearly immediate. Can this apparent discrepancy be resolved? Explain.

6. What learning theorist discussed in this volume do you feel would support Brownell's view of learning as progressing "by transversing a series of stages in thinking"? If you were to accept this view of learning, how would you teach solving equations of the form $ax + b = c$? (or the division algorithm?)

7. Some would criticize Brownell's stages as supplying the learner with "crutches" which are devices which must be unlearned later. What evidence can you give from the classroom supporting this criticism of Brownell. Where do you stand on the issue? Illustrate and defend your position.

8. What about "Practice makes Perfect"? Is the phrase defensible or is it a dangerous saying? Do you agree with Brownell? Illustrate with an example from the classroom.

9. Choose a modern textbook for elementary school which deals with the teaching of problems similar to "42 plus 27." How

do the strategies suggested by the book stack up to Brownell's recommendations? Explain.

10. Contrast Brownell's description of connection-forming with Cronbach's discussion of prompting.

11. John has taken a quiz concerned with the multiplication of integers. He has done poorly. List the types of diagnostic evidence that would indicate practice as the best treatment for his difficulty.

Jennifer

Robert B. Davis
Rhonda Greenstein

Teachers are, we believe, all too aware that most "educational research" does not seem to help them very much in their daily work in the classroom. There are probably several reasons for this: often the research does not address itself to the questions that are of greatest concern to the teacher; the controlled studies often operate in a setting and under ground rules which are quite different from ordinary school situations; the cues which mean the most to teachers—a look of bewilderment, a look of sudden insight, a seeming sense of hopelessness, and other matters of facial expression, vocal inflection, and body posture—are often largely neglected or eliminated in "antiseptic" research settings.

One possible response to this situation is to undertake some research studies of a more personal or "clinical" sort, in which human interactions are as natural as possible, and are observed carefully and interpreted shrewdly. We present here one such "case study" of teacher-child interaction.

Background: the child was a rather alert 8 year old, nominally a third grader, but actually attending a non-grade (or "family plan") classroom in which one teacher, in one room, dealt simultaneously with 6 year olds, 7 year olds, and 8 year olds. The instructional style was oriented (as it had to be, in the face of a deliberately very heterogeneous group of children) heavily toward individualized instruction and small-group work. The school was better than average, was racially integrated, and was in a neighborhood which qualified for federal assistance because of poverty conditions. The classroom teacher would probably have to be rated as considerably better than average.

Even with a strong orientation toward individualization, the pressures of 30 children in one room ordinarily precluded the degree of individualization which we shall describe presently. In regular classroom encounters the teacher had become aware that the girl, Jennifer, was somewhat more capable than the curriculum called for, and that Jennifer was working on problems in "long" division, but with very limited success.

At this point it became possible, because of the use of parent volunteers, teacher apprentices, university student volunteers, and paraprofessionals, to achieve a much higher degree of individualization. We now describe a session in which one highly experienced teacher worked with Jennifer; the session was observed by three adult observers, but was not recorded on videotape or audiotape. The following record is based upon written notes; dialogue is only approximate, because of the limitations of written recording techniques. By far the most important

Robert B. Davis and Rhonda Greenstein, "Jennifer," *New York State Mathematics Teachers' Journal*, Vol. XIX, **3**, June 1969, pp. 94-103. Reprinted with permission.

aspect of this session is this: the point where a child *seems* to have difficulty is not necessarily (and, in our experience, is not *usually*) the place where he is *really* having trouble. Indeed, his *real* difficulties probably lie in the area of earlier concepts which he has failed to understand. An analogy with medical diagnosis immediately suggests itself—or with dentistry. If your tooth hurts *today,* it is probably due to some problem which was well under way last week, and most likely even earlier than that.

The presenting symptom: the classroom teacher had recognized that Jennifer experienced difficulty in using (as the curriculum suggested) a "long division" approach to problems such as

$$8\,\overline{\smash{\big)}\,4808}$$

and, indeed, this specific problem was suggested by the classroom teacher as a starting point for the special "individualization" teacher (who had not previously met Jennifer). The "individualization teacher" conducted the session which we now report.

The *individualization teacher,* wishing to get further insight into Jennifer's level of understanding, began with the recommended problem

$$8\,\overline{\smash{\big)}\,4808}$$

and asked Jennifer what she knew about it.

Jennifer, in response, replied that "8 doesn't go into 4, so you have to say that 8 goes into 48, 6 times" and wrote

$$8\,\overline{\smash{\big)}\,\overset{6}{4808}}$$

[Jennifer's strategy seems clear: since she has "used the entire 48," she has written the 6 so that it is centered over the 48.]

The *individualization teacher* responded by suggesting that, since indeed the 48 had been "all used up," it was better to write the 6 over the right-hand end of the 48.

Jennifer changed her response to

$$8\,\overline{\smash{\big)}\,\overset{\;\;6}{4808}}$$

Individualization teacher: "All right. What do you do next?"

Jennifer: "The zero doesn't count for anything, so you say '8 goes into 8 once.'" She wrote:

$$8\,\overline{\smash{\big)}\,\overset{\;\;6\;1}{4808}}$$

Here, we feel, is the first crucial decision point for a teacher. We can think of 4 general lines of attack:

i) We can be quite direct, and tell Jennifer to write a "zero" over the "zero" in "4808." Jennifer was a well-behaved child and would surely comply. Conceivably, after a few further similar examples, Jennifer would be able to anticipate the teacher's response, and even to imitate it. [We would not select this approach ourselves; we fear that we might end up treating the symptoms by prescribing aspirin, when we ought to diagnose and treat the disease, by appropriate surgery. Nonetheless, we do not consider this approach hopeless. Indeed, *there are exceedingly few strategies that will not succeed in the hands of a highly resourceful teacher who really believes in them.*]

ii) One might try to "explain" this problem, by using "expanded notation" such as

4 thousand + 8 hundred + 0 tens + 8 ones,

or by some other paper-and-pencil or verbal exposition. The individualization teacher rejected this alternative: thus far, Jennifer had been alert and actively cooperative, but the individualization teacher felt that Jennifer would "tune-out" any paper-and-pencil or verbal "explanation" of this type. Even if she tolerated it, Jennifer might be picking up words she did not understand, just as one can sing "Adeste, fidelis, . . ." without being sure what "adeste" means. This might be another form of aspirin therapy, avoiding the disaster today by merely postponing it.

iii) One could conclude (surely with good reason) that Jennifer was none too clear on "ten" "hundred" and "thousand," and could use concrete materials such as Zoltan Dienes' well-known MAB blocks,[1] or the Beryl Cochran-Jane Drucker "beans and tongue depressors"[2] materials. This would seem to be an eminently reasonable response, and this same individualization teacher has used precisely this approach with other children quite similar to Jennifer; the results seem to have been satisfactory. [Nonetheless, the teacher, suspecting that he did not yet fully understand the nature of Jennifer's difficulties, chose instead to pursue further probing diagnosis before embarking on a program of therapy. Therefore he selected option number iv:]

iv) One could pursue further back the question of what Jennifer did, and did not, understand. This requires some technique for probing. The individualization teacher chose this option, and used the following techniques:

a) The use of contrasting examples;
b) A search for simpler, similar problems;
c) A search for a "bed-rock" foundation which Jennifer clearly understood;
d) After finding this "bed-rock" foundation in Jennifer's past, retracing one's steps back to the problems of the present, in such a way as to build carefully on the best available foundation at each stage;
e) Frequent recourse to letting Jennifer carry on according to her own desires;
f) Introduction of examples specifically intended to challenge any of Jennifer's methods which were, in fact, inadequate.

---

[1] Available from: Herder and Herder, Inc., 232 Madison Avenue, New York, New York 10016.

[2] Cf. *A Modern Mathematics Program As It Pertains to the Interrelationship of Mathematical Content, Teaching Methods and Classroom Atmosphere.* The Madison Project. Report submitted to the Commissioner of Education, U. S. Department of Health, Education and Welfare, 1967. Volume II.

Here is how the session continued:
Since Jennifer had written

$$\begin{array}{r} 6\ 1 \\ 8\ \overline{)\,4808} \end{array}$$

the teacher (using what we would call "technique a" in the preceding list) asked:
Teacher: Would these two problems be the same:

$$8\ \overline{)\,4808} \qquad 8\ \overline{)\,488}$$

Jennifer: Yes
[This response took the teacher quite by surprise. Note, however, that Jennifer is being entirely consistent within her stipulated point of view: since "0 doesn't make any difference," Jennifer will get the same answer in both cases. Indeed, she worked each example, and in each case obtained the "answer" 61.]

This begins to show the main trend of the session, and the main theme of this report: *the trouble lies further back than you might, at first, suspect!*

To begin with, we believed Jennifer had difficulty with long division. It now looks as if Jennifer does not understand place-value numerals, and in particular has no concept of the size of 4,808 vs. 488. The teacher probes further:
Teacher: How about this example:

$$8\ \overline{)\,800}\ ?$$

Jennifer: 8 goes into 8 once:

$$\begin{array}{r} 1 \\ 8\ \overline{)\,800} \end{array}$$

Teacher [still probing further into Jennifer's background, in hopeful search of that solid bed-rock foundation]: Which of these is the largest:

$$8 \qquad 80 \qquad 800\ ?$$

Jennifer [in whom consistency is a great virtue]: They're all the same. Zero doesn't make any difference.

By now the teacher was convinced that numerals were, to Jennifer, mainly meaningless marks on paper; consequently he made an all-out plunge for *some* numbers that *did* have meaning. Within the teacher's experience, even 3-year-olds have some feeling for *ages of young children,* and usually know that 4-year-olds are bigger than 3-year-olds. Hence he proceeded as follows:
Teacher: How old are you?
Jennifer: Eight. [That didn't help much, since for the teacher's purpose it was necessary to get some zeros into the act.]
Teacher: [Trying again] Do you have any older brothers or sisters?
Jennifer: A brother [whose age turned out to be 9 years old] and a sister [whose age turned out to be 11 years old].

The Teacher had Jennifer write "9" and "11" [unfortunately, still no zeroes!].

Teacher: Do you have any friends who are older than your brother, but younger than your sister?

Jennifer [smiling triumphantly]: I'm younger than my sister, and my father is older than my brother. [This is a very common error in children when asked questions of this type.[3]]

Teacher: You used *two* people, yourself and your father. I want you to tell me *one* person who is *both* older than your brother, and younger than your sister.

Jennifer [with what we regarded as a more genuine facial expression of understanding]: Oh! Somebody who was ten! [However, Jennifer did not know anyone who was 10.]

Teacher: Can you write "ten"?

Jennifer could, and did.

Teacher: Which is larger, [writing on paper]

$$1 \qquad \text{or} \qquad 10 \ ?$$

Jennifer: They're both the same, because zero doesn't make any difference. [Suddenly a *real* expression of comprehension lit up her face:] Oh, that's *one,* and that's *ten!*

[The teacher felt that, at long last, he had found the solid bed-rock on which he could build. Jennifer may have been confused, previously, by the "insignificant" zero in the written symbols "1" and "10," *but she really did know how much "one" was, and she really did know how much "ten" was!* The phenomenon that some call "cognitive dissonance" now compelled her to carry through an agonizing reappraisal of the written symbols "1" and "10"—and she did in fact carry it through, with considerable success.]

At this point Jennifer responded in a fashion very common among children, as David Page has pointed out very accurately: with no prompting from the teacher, Jennifer eagerly and immediately turned to the *preceding* problem—we might call this the "solecism effect"—and, realizing that she must have answered incorrectly there, she now straightened out the difference between the written symbols:

$$8 \qquad 80 \qquad 800$$

Brief report on what followed: the teacher returned to the problem of 488 and 4808, and asked Jennifer to interpret these numbers as money. Jennifer chose to do this as $4.88 vs. $48.08, and thought of things (such as a pair of shoes) that might cost about $4.88, and things (she suggested "an expensive gown-type dress") that might cost $48.08.

The teacher pursued money somewhat further (e.g., Jennifer correctly suggested that one of the things that might cost about $4,808, would be an automobile), and then began a second backward probing into Jennifer's comprehension of earlier ideas, this time in relation to the ideas of

---

[3]We are inclined to guess that schools often spend far too much time "establishing" the ideas of *more* and *less, older* and *younger*—ideas which most children know all too well without our help—with the result that one gets a kind of conditioned response, which is what Jennifer was giving us here, "whenever asked, say *more* or *less* or *older* or *younger*, as the case may be!"

addition, multiplication, subtraction, and *division.*[4] We do not reproduce this here, except to say that Jennifer showed corresponding weaknesses in this area as well. The teacher used the British approach of making a picture such as

$$\begin{array}{cc} \times & \times \\ \times & \times \\ \times & \times \end{array}$$

and asking Jennifer to write *as many arithmetic stories as possible* that might describe this picture. Jennifer (after some struggling and more back-tracing of past journeys) wrote:

$$3 + 3 = 6$$
$$2 \times 3 = 6$$
$$6 - 3 = 3$$
$$\tfrac{1}{2} \text{ of } 6 \text{ is } 3$$

Let us leave Jennifer, for the moment, and see what this session may suggest:

1. It would surely have been possible to patch up Jennifer's difficulty with

$$8\,\overline{\smash{)}4808}$$

by, in effect, telling her what to do. A medical or dental analogy is nearly irresistible—for a toothache, take aspirin, and the pain will go away today, although you may lose the tooth tomorrow—and in our view this medical analogy may be all too accurate. Perhaps, all too often, when a child has trouble with today's lesson, we need to search out, and try to correct, misunderstandings from last week, or last year, or even several years back. Perhaps, all too often, we fail to do this, and thereby achieve a painless today at the cost of a toothless tomorrow.

2. Jennifer's example is not unusual within our experience. It would be closer to the truth to say that her case is typical. (As one other instance, we have recently worked with a 1st grade girl, in February, who was seemingly "having difficulty with subtraction"—for example, given the problem

$$3 + \square = 5$$

she would write $8 \rightarrow \square : 3 + \boxed{8} = 5.$

In attempting to study this case we used balances, toy people divided among two rooms of a house, etc., and—to our surprise—discovered that

---

[4]For example, when asked to interpret "2×3" in terms of addition, Jennifer could not come up with the idea "$3 + 3$" (although most children this age with whom we have worked produce this immediately). Instead, Jennifer suggested the picture

$$\begin{array}{cc} \times & \times \\ \times & \times \\ \times & \end{array}$$

and the "story" $2 + 3 = 5$, although she felt that both of these seemed somehow to be wrong. When asked to say what multiplication meant, she could not think of anything at all.

the girl could not count out six pebbles and actually get six. Yet on paper and pencil tests such as

$$3 + 2 =$$

she gave mainly correct answers. Almost certainly she got through a large number of "addition facts" *merely by memorizing meaningless words and meaningless phrases.* Apparently this prodigious feat of memorization carried her until February of first grade, but by then this system was strained to its limit, and it was incapable of carrying her any further. Was there any hope at all for her aside from going back to what "2" or "3" mean, counting out two pebbles, counting out 5 pebbles, and so on, and thereafter retracing the year's work, attempting to build more securely? We find many such children, indeed—and isn't this rather like undiscovered cases of tuberculosis or diabetes? We need to seek these cases out, diagnose them correctly, and treat the disease instead of the symptoms.)

3. In our view this is related to what we have called the "superficial verbal problem": all too often children can say or write something that appears to be correct—like singing "Adeste, fidelis . . ." *without having any idea of the meaning of what they have said or what they have written.* All they know is that this seems to be what the teacher wants, so you'd better do it this way . . .

4. If this difficulty is in fact as widespread as we believe, many modifications in school learning tasks and in school testing procedures are needed to prevent such cases wherever we can, and to discover the victims whenever prevention has failed. *One device that we strongly recommend is the widespread use of manipulatable materials.* Most classes that we have studied treat mathematics mainly as a paper-and-pencil or "conversational" matter, and seem to us to make far too little use of manipulating actual physical materials—which can often be as simple as counting out pebbles or counting out drinking straws. Tests, also are too often limited to paper and pencil.

5. There are devices other than physical materials. One is the kind of searching diagnostic questioning procedure that the teacher used with Jennifer. Another is the use of "environmental" experience (also used with Jennifer), such as consideration of the ages of different children, the cost of items with which the children are familiar—and indeed, the study of currency itself (although this is actually best undertaken by the use of genuine coins, and may therefore fall under the earlier category of "use of manipulatable physical materials").

6. We believe there are even more far-reaching implications. If such cases are in fact common, should we remove from the teacher the constraint of a rigid curriculum with rigid grade-level expectations? Under "curriculum" pressure the teacher may not feel that she can afford the time to go all the way back to counting out drinking straws (or whatever the individual case may require), and may feel compelled to resort to the band-aid strategy of "aspirin today and total failure sometime next year." With this curriculum pressure removed, the teacher might feel free to try to find out where Jennifer's weaknesses *really* lie,

and to go as far back as possible in order to find a good solid bedrock of understanding upon which to build.

Such a relaxation of grade level expectancies, if wisely used, would in our opinion help a large number of children. But it would demand that we ask ourselves whether we want to graduate from grade six children who speak correct language at the fourth grade level but do not understand their own words, even at a first grade level. Or would we prefer to back-track, take the time to do a more thorough job, and possibly graduate from grade six a child who was slightly less articulate, whose formalized statements put him perhaps at the third grade level—*but who actually knows, at the third grade level, what he is talking about?*

Our belief is that he who knows not, and knows not that he knows not, would be better off making sense to himself and everyone else even at the price of presenting a less sophisticated facade, albeit a more genuine one.

It would be pleasant to conclude by reporting that Jennifer now knows all about division, including

$$8\,\overline{)\,4808}$$

(and even $8\,\overline{)\,4907}$),

but in fact our work with her continues. What is genuinely encouraging is that Jennifer is demonstrably learning more, that we believe her understanding to be at least reasonably genuine, and that we believe that adults and Jennifer are understanding one another better.

We are aware that this, and similar instances, pose several questions to which we do not know the answers. We will be grateful for suggestions from readers.

i) "Long division" for children is in fact usually a paper-and-pencil algorithm. But is this necessarily wrong? Can we legitimately allow our development of what is essentially a *linguistic* structure to proceed far ahead of any "content" or "meaning"?

(We asked a 10-year old girl how *she* would explain the algorithm to Jennifer, and she suggested that first she would do a problem in addition:

"You don't write $\begin{array}{r} 50 \\ +\,10 \\ \hline 6 \end{array}$ but instead you write $\begin{array}{r} 50 \\ +\,10 \\ \hline 60 \end{array}$

and it's the same way with

$$8\,\overline{)\,4808}^{\,601}\,.")$$

ii) Can children really derive an adequate *meaning* (as opposed to adequate *performance* of the algorithm) from the now-common "subtractive" algorithm?

$$
\begin{array}{r|r}
6\,/\,240 & 10 \\
\underline{60} & \\
180 & \\
\underline{60} & 10 \\
120 & \\
\underline{60} & 10 \\
60 & \\
\underline{60} & 10 \\
\hline
0 & 40
\end{array}
$$

iii) Would the approach above (question ii) be further enhanced by a careful sequence of questions that led the child to "discover"—as a "shortcut"—the usual treatment of zero?

iv) Do the difficulties of long division justify raising the question of why we do long division in elementary school, anyhow?

v) Would we be better off if we used MAB blocks to establish a linguistic system based on "longs," "flats," etc., and then carried it over to "tens," "hundreds," etc.?

vi) Would it be better to develop the initial algorithm in base 3, so that the numbers would not be so large and unmanageable (as 4,808 is)? (One would then translate the algorithm to base 10 numerals.)

vii) Does anyone have a better idea?

Exercises: Jennifer

1. When Jennifer was asked about 1 and 10 she was tempted to say that they were the same, yet she "knew" they were different. How could you tell what a child knew without requiring a verbal response?

2. Consider the girl who could not count out six pebbles and actually get six, but on paper and pencil tests such as $3 + 2 = ?$ she gave mainly correct answers. Describe and explain this situation using Gagné's eight types of learning.

3. What cautions should be considered by the teacher who is constructing a learning hierarchy according to Gagné's procedures?

4. Davis suggests the use of manipulative materials in this elementary school class. What manipulative materials might be useful in algebra? in geometry?

5. Why are "environmental" experiences or examples from social situations important in the classroom? Does this importance change as students move to higher grade levels?

6. This article focuses on one student? How does it relate to individual differences?

7. Davis makes a distinction between being able to perform an algorithm and knowing the meaning of the algorithm. He indicates that one may perform without knowing the meaning. Can one know the meaning without being able to perform the algorithm?

8. Examine the questions posed by Davis, i-vi. Suggest partial answers.

Piaget Interviews

The following are actual interviews conducted by Piaget and his researchers. We give them here to allow you to see how Piaget's research was conducted and to perhaps do some sample interviews of your own. The notation John (5; 6) means that the child's name was John and he was 5 years, 6 months old.

. . . Six little bottles (about one inch high, of the kind used in dolls' games) are put on the table, and the child is shown a set of glasses on a tray: "Look at these little bottles. What shall we need if we want to drink?—*Glasses.*—Well, there they are. Take off the tray just enough glasses, the same number as there are bottles, one for each bottle." The child himself makes the correspondence, putting one glass in front of each bottle. If he takes too many or too few, he is asked: "Do you think they're the same?" until it is clear that he can do no more. Mistakes occur in fact only with children of the first stage (4-5 years). The correspondence can be made easier by getting the child to empty the bottles into the glasses, each bottle just filling one glass. Once the correspondence is established, the six glasses are grouped together and the child is again asked: "Are there as many glasses as bottles?" If he says "no," he is then asked: "Where are there more?" and "Why are there more there?" The glasses are then rearranged in a row and the bottles grouped together, the questions being repeated each time.

Bon (4;0): "Look at all these little bottles. What shall we need if we want to drink?—*Some glasses.*—Well, there are a lot here (putting them on the table). Now put out enough glasses for the bottles, just one for each.—(He took the 12 glasses, but put them close together, so that the 6 bottles made a rather longer row.)—Where are there more?—*There* (the bottles).—Well then, put one glass for each bottle.—(He made the 12 glasses into a row the same length as that of the 6 bottles.)—Are they the same?—*Yes.*—(The bottles were then put further apart.) Is there the same number of glasses and bottles?—*Yes* (but he spread out the glasses a little more.)—(The bottles were then put still further apart.)—*There are only a few here* (the 12 glasses), *and there* (the 6 bottles) *there are a lot.*"

Car (5;2): "Arrange them so that each bottle has its glass.—(He had taken all the glasses, so he removed some and left 5. He tried to make these correspond to the 6 bottles by spacing them out so as to make a row the same length.) Is there the same number of glasses and bottles?—*Yes.*—Exactly?—*Yes.*—(The 6 bottles were then moved closer together so that the two rows were no longer the same length.) Are they the same?—*No.*—Why?—*There aren't many bottles.*—Are there more glasses or more bottles?—*More glasses* (pushing them a little closer together.)—Is there the same number of glasses and bottles now?—*Yes.*—Why did you do that?—*Because that makes them less.*"

Hoc (4;3): "Look, imagine that these are bottles in a café. You are the

---

From Jean Piaget, *The Child's Conception of Number* (New York: Humanities Press, 1952); (London: Routledge & Kegan Paul Ltd.). Reprinted with permission.

waiter, and you have to take some glasses out of the cupboard. Each bottle must have a glass." He put one glass opposite each bottle and ignored the other glasses. "Is there the same number?—*Yes.*—(The bottles were then grouped together.) Is there the same number of glasses and bottles?—*No.*—Where are there more?—*There are more glasses.*" The bottles were put back, one opposite each glass, and the glasses were then grouped together. "Is there the same number of glasses and bottles?—*No.*—Where are there more?—*More bottles.*—Why are there more bottles?—*Just because.*"

Gal (5;1) made 6 glasses correspond to 6 bottles. The glasses were then grouped together: "Is there the same number of glasses and bottles—*No, it's bigger there* (the bottles) *and smaller here* (the glasses).—(The bottles were then grouped together and the glasses spread out.)—*Now there are more glasses.*—Why?—*Because the bottles are close together and the glasses are all spread out.*—Count the glasses.—*1, 2, . . . 6.*—Count the bottles.—*1, 2, . . . 6.*—They're the same then?—*Yes.*— What made you say they weren't the same?—*It was because the bottles are very small.*"

Fu (5;9) poured the contents of the 6 bottles into 6 glasses and put the glasses in front of the empty bottles. "Is there the same number of bottles and glasses?—*Yes.*—(The bottles were grouped together in front of the glasses.) Are they the same?—*No.*—Where are there more?—*There are more glasses.*—(The reverse process then took place.) And now?—*There are more bottles.*—What must we do to have the same number?—*We must spread out the glasses like this, no, we'll need some more glasses.*"

Pel (5;6) began by putting 5 glasses opposite 6 bottles, then added one glass: "Are they the same?—*Yes.*—And now (grouping the glasses together)?—*Yes, it's the same number of glasses.*—Why?—*That hasn't changed anything.*—And if they're like that (grouping the bottles together and spacing out the glasses)?—*Yes, it's the same.*"

Lau (6;2) made 6 glasses correspond to 6 bottles. The glasses were then grouped together: "Are they still the same?—*Yes, it's the same number of glasses. You've only put them close together, but it's still the same number.*—And now, are there more bottles (grouped) or more glasses (spaced out)?—*They're still the same. You've only put the bottles close together.*"

Angular Measurement*

The subject is shown a drawing of two supplementary angles ADC, CDB, and is asked to make another drawing exactly similar. He is not

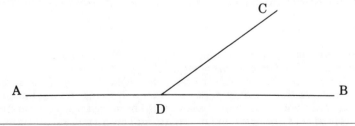

---

*From Jean Piaget with Barbel Inhelder, and Alina Szeminska, *The Child's Conception of Geometry* (New York: Basic Books, Inc., 1960), pp. 173-178, 180-183. Reprinted with permission.

permitted to look at the model while he is drawing, but he may study and measure it as often as he wishes while not actually engaged on his own drawing. This requirement is met quite simply by having the model behind the subject. The latter is provided with rulers, strips of paper, string, cardboard triangles, compasses, etc., all of which may be used in measuring. The enquiry is repeated three times in all, the drawings being similar in each case.

Jea (6;5) starts by drawing a horizontal line AB together with an upright CD, almost perpendicular to and bisecting AB, all by eye: "Is that right?—*No, it's not good.*—If there's anything you want, you can use whatever you find on the table.—*No, I don't need anything* (he produces another drawing, this time with the angle too acute).—Wouldn't you get it better with the ruler?—*No, I can draw a straight line without a ruler.*—What about measuring?—No, it's difficult with a ruler."

Nid (6;6) draws a horizontal line AB and a line CD which joins AB at a very acute angle and nearly at point A. "Is it right?—*Not quite* (he starts afresh, making CD shorter, but without troubling about its slope).—(A second model.) What about this one?—(He goes about it in a similar manner producing a drawing with angle CDB too obtuse.)"

Dre (7;3) after sketching two lines with his hand, takes the pencil and draws AB and CD without measuring them. "Is it the same?— *Yes* (but on careful inspection he changes his mind and makes a fresh start. This time he measures AB with thumb and forefinger).—I've given you anything you may need to measure.—*I don't need anything.* (Nevertheless he takes a ruler and measures AB, transferring the measurement to his own drawing. But CD is drawn without any measurement.) *I'm doing it right and still it won't work.*"

Ire (7;4) looks at the models saying: "*I need a ruler to do those, so as to draw the lines right* (as indeed he does, but without measuring them. He then finds his copy is incorrect and therefore measures AB and CD but continues to ignore the position of point D). *That's funny. I've measured it right, but it still won't work!*—Why not? Isn't there something else you might measure?—*No. I've measured that* (AB) *and that* (CD) *and that's all there is* (!)."

Den (7;7) first draws the figure entirely by eye, then measures AB and CD: "Is that right?—*No, I measured it wrong* (making a second attempt).—All right now?—*No, but I don't know why not. I measured it very carefully.*—Isn't there something else you could measure?—*No, there are only two lines, and I've measured them both.*"

Hel (7;10) measures AB and inserts D by inspection. He then measures DC and moves the ruler to his own drawing. "*It won't work.*—Why not?—*Because I carried the ruler wrong.* (Measures AD.)—Why do you measure that?—*Because I've got to know where to begin this line* (DC. He draws this line but is still not satisfied with its slope). *No good.* (He tries again, seeking not to alter the slope of his ruler.) *I've measured it wrong and I went wrong when moving my ruler.*"

Rev (8;3) measures AB and AD "*to get it more right*". He then measures CD and transfers the measurement to his own drawing, trying hard to keep the ruler at a constant slope. "It isn't right because I was very careful about the slope of my ruler when I moved it, but I didn't measure it right." Rev seems to think he was more accurate in transferring the slope of CD than he was in measuring it!

Gal (8;6) begins without measuring but is dissatisfied with the result. He therefore measures AB and CD and produces another drawing: "Is that right?—*No, mine is straight and it should slope that way more.*" He measures AD and plots point D, then draws DC trying to preserve the required slope. Again he studies his drawing and says: "*I must have the distance from there to there* (BC)." He measures AC and CB and alters DC, having found a unique position for point C.

Lau (9;0) measures AB, marking D on his ruler as he does so, then, after drawing the base line ADB he draws the beginning of CD on the ruler itself as well as a line at right angles to it. However, when he finds that the angle he draws is too sharp, he measures the vertical distance CK (being the perpendicular from C on AB), and so finds a unique position for point C by combining this measurement with DC.

Rol (10;4) measures AB, then studies DC saying: "*I can't copy that line because it isn't straight* (i.e. vertical)." He then measures AC and DC to determine point C, and checks his finding by measuring CK. "Is that right?—*Yes, but I'll do it again* (he makes a second copy, this time measuring only CK and omitting AC). *It's easier this way, and I can do it quite quickly.*"

Jac (10;8) measures AB and AD and a line perpendicular to AB drawn from point B and ending at K". He then measures CK" (parallel with AB) and so draws line DC. "Why did you do it that way?—*To get this point* (C) *right.*—Is there no other way of doing it?—*Yes, I could do the same with either side* (i.e. a perpendicular on A) *but it's exactly the same.*"

Mir (10;8) measures AB and AD followed by DC but instead of drawing DC he plots a provisional point at C and then draws a line from C perpendicular to AB (CK). He measures this line and then finds where the vertical distance coincides with DC. "Why did you do it that way?—*I didn't know where to draw this line* (DC). *I knew where to begin* (D) *but not where to end. This way I can tell where this point* (C) *comes on this* (CK), *so I can do it quite easily.*"

The Equality of Angles of Incidence and Reflection and the Operations of Reciprocal Implication*

Our aim . . . is not a systematic study of the concept of the equality of two angles. Actually, we already know how the concept is constructed: that it is first acquired at the level of concrete operations.[1] But it is precisely the fact that the concept is already so well known by the time the formal level (stage III) is reached that makes the reasoning process involved in the discovery of the equality between the angles of incidence

---

*From Barbel Inhelder and Jean Piaget, *The Growth of Logical Thinking from Childhood to Adolescence* (New York: Basic Books, Inc., 1958), *passim.* Reprinted with permission. With the collaboration of H. Aebli, former research assistant, Laboratory of Psychology, Science Faculty, University of Geneva, professor, École normale supérieure, Zurich; L. Müller, former research assistant, Institut des Sciences de l'Éducation, University of Geneva; and M. Golay-Barraud, student, Institut des Sciences de l'Éducation.

[1] See Piaget and Inhelder, *The Child's Conception of Space* (London: Routledge & Kegan Paul, 1956), chapter XII, and Piaget, Inhelder, and Szeminska, *La géométrie spontanée de l'enfant,* chapter VIII. (Not transl.)

and reflection so instructive. One of the aims of this study, then, is to isolate the operational mechanisms involved in the formal reasoning process itself, when this reasoning rests on notions already constructed at the concrete level.

The experimental apparatus consists of a kind of billiard game. Balls are launched with a tubular spring device that can be pivoted and aimed in various directions around a fixed point. The ball is shot against a projection wall and rebounds to the interior of the apparatus. A target is placed successively at different points, and subjects are simply asked to aim at it. Afterwards, they report what they observed.

Dan (5;2) succeeds at first: *"I think it works because it's in the same direction."* He adjusts the plunger by himself, but proceeds by empirical trial-and-error. Then he asks spontaneously: *"Why do you have to turn the plunger sometimes? . . . No, you have to put it there* [he fails]. *If it could be pushed a little further"* [he does this and succeeds]. But, although he knows how to control the rebounds successfully, Dan has no idea that they are made up of angles: the curve he describes with his finger is not tangent to the wall; he takes into account the starting point and the goal but not the rebound points.

Nan (5;5), on the other hand, is astonished by the detour made by the ball which first touches the walls. *"It always goes over there."* But he does not succeed in adjusting his aim: *"Oh, it always goes there. . . . it will work later."*

Per (6;6), in contrast, in spite of his age, resorts to the curvilinear model: *"It goes there and it turns the other way"* [gesture indicating a curve].

Vir (7;7) succeeds after several attempts. He points out and then draws trajectories with two distinct rectilinear segments, saying: *"To aim more to the left, you have to turn* [the plunger] *to the left."*

Truf (7;10): *"I know about where it will go";* in fact, he shows by his gestures that he realizes that the angle of rebound is extremely acute when the plunger is raised and extremely obtuse when it is lowered. Thus, he shows us that he has a vague global intuition of the equality between the angles of incidence and reflection. But he does not make it explicit, since he fails to divide the total angle indicated by his gesture into two equal angles.

Desi (8;2): *"The ball always goes higher when the plunger is higher."* Then: *"The ball will go there* [further] *because the plunger is tilted more; I put my eyes high up* [= I pinpoint the rebound point] *and from the rubber* [= the rubber band attached to the wall on which the ball rebounds] *I look at the round pieces"* [= the disks serving as goals].

Kar (9;6): *"The more I move the plunger this way* [to the left—i.e., oriented upwards], *the more the ball will go like that* [extremely acute angle], *and the more I put it like this* [inclined to the right], *the more the ball will go like that"* [increasingly obtuse angle]. Kar reaches the point of discovery that the ball returns to the starting point when the plunger is *"straight"*—i.e., perpendicular to the rebound wall.

Ulm (9;8): *"As you push the plunger up, the ball goes more and more like that* [acute angle], *and the more I put it like that* [inclined to the right], *the more the ball will go like that"* [obtuse angle].—"But, tell us more about what you are looking at."—*"I am still looking at that* [the goal], *and that's all, because it turns with the plunger"* [—because the direction of the path between the rebound and the goal changes with the inclination of the plunger].

Bon (14;8) first invokes the launching force, then realizes that the trajectories are the same whether the balls are shot hard or soft. Next he invokes the role of the *"distances, how you have to place the rod."* Then he establishes concrete correspondences in the same way as the stage II subjects: *"It's the position of the lever* [of the plunger]: *the more you raise the target, the more you raise it here"* [the lever]. He uses a ruler to mark the trajectory of the ball between the rebound point and the target in such a way as to verify its correspondence with the orientation of the plunger. Then he hypothesizes that the angle is always a right angle: *"It has to make a right angle with the lever."* But after several trials he concludes: *"No, above* [= when the plunger is straightened] *it won't work."*—"It isn't ever a right angle?"—*"Yes, that's correct for one position."*—"And without that?"—*"When you turn, one should be smaller, the other larger. Ah! They are equal"* [he points out the angles of incidence and reflection].

Mul (14;3) begins with a series of correspondences: *"I was here and it went in this direction,"* etc.; *"You change the angle to see how it goes."* By systematically diminishing the total angle, he discovers the fundamental proposition: *"If I shoot it straight, at a right angle* [i.e., when the plunger is perpendicular to the buffer], *it will come right back."* Then he inclines the plunger progressively, according to the angles a_1, β_1, γ_1, etc., and ascertains that, as these angles increase, their complementaries a_1', β_1', γ_1' decrease [a', β' standing for the angles included between the plunger and the buffer]: *"The smaller you make the angle here* [a_1', β_1', etc.], *the larger the angle there"* [a_1, β_1, γ_1]. Then he perceives the equality which he had been seeking from the time he understood that, in the case where the plunger is perpendicular, the ball returns to its starting point. *"This angle* [a_1'] *is the same as that one* [a_2']: *you have to make it parallel to that one* [a_2']. *I am going to see* [he checks for several different angles]. *Yes, I think that's it. You have to carry over exactly that angle"* [the complementaries a_1', and a_2', etc.].

Rev (15;4): *"It's a right angle* [several trials]. *No, this slant has to be the same as that one."* When there are chance misses due to the apparatus, he says, *"I didn't move; the gadget isn't fair."*

The Projection of Shadows*

In addition to the usual problem of the formal operations needed to establish the table of possibilities that allows the discovery and verification of a law, the present research raises a question about the formal operational schema relative to proportionality. But we are dealing with a new type of proportionality. Whereas the proportions in the problems of the balance and of hauling a weight on an inclined plane derive from a model of physical equilibrium, the proportions we shall study in connection with the projection of shadows are of an essentially geometrical nature. They denote relationships between distances and diameters in a physical phenomenon that can be explained in terms of simple projective geometry.

The law to be discovered in this experiment is extremely simple. Rings of varying diameters are placed between a light source and a screen. The

---

*With the collaboration of Vinh Bang, research assistant, Institut des Sciences de l'Éducation; B. Reymond-Rivier, research assistant, Institut des Sciences de l'Éducation; and F. Marchand.

size of their shadows is directly proportional to the diameters and inversely proportional to the distance between them and the light source. Specifically, we ask the subject to find two shadows which cover each other exactly, using two unequal rings. To do so he need only place the larger one further from the light, in proportion to its size, and there will be compensation between distances and diameters.

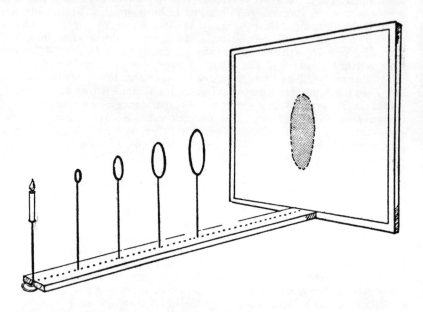

The projection of shadows involves a baseboard, a screen attached to one end of this, a light source, and four rings of varying diameters. The light source and the rings can be moved along the baseboard. The subject is asked to produce two shadows of the same size, using different-sized rings.

Pel (7;10) predicts correctly that a ring 10 cm. in diameter will produce a larger shadow than a ring of 5 cm., etc. "If I move it to this side, where will the shadow be?"—*"There"* [accurate]. "Does it stay the same size or does it get bigger or smaller?"—*"It's the same."*

Mand (9;6): *"As it advances* [toward the light], *it* [the shadow] *always becomes bigger, because when it is closer* [to the screen] *it gets smaller, and when it's further away* [from the screen] *it gets bigger."*

Oli (10;2) puts the 5 cm. ring at 10 and the 10 cm. at 19, then 15 at 38 and 20 at 50. "Why did you put them that way?"—*"Because with those* [5 and 10 diameters] *it's bigger* [because closer to the light] *and those* [15 and 20 diameters] *get smaller."*

Che (12;8) measures the rings and finds that their diameters differ by 5 cm. He concludes that one must *"find a distance between them which is a multiple of 5."* He places them correctly in proportion to size.

Duc (12;1), after having placed the 20 cm. ring at 83, says: *"Now you have to count from here to there* [to the light source] *and divide by 4."*

Then he puts the 5 cm. ring at 21, the 15 cm. at 61, but the 10 cm. at 51. "And if I only give you three rings?" [5, 10, and 15].—*"You have to count from the largest and divide by three."*

Mic (14;6): *"Since the diameters all have regular differences, the differences between the distances have to be the same."* Then he places the rings of 5, 10, 15, and 20 cm. at distances of 8, 16, 24 and 32 cm., respectively; next, he takes another arbitrary distance and finds the proportion in the same way: *"The distances have to have the same relation to each other as the rings."*

Mart (16;2) begins getting the rings to coincide: *"You have to put the largest the furthest away, and the ratio between the diameters of the rings and the distances has to be the same."* He is successful in discovering the proportion.

Exercises: Piaget Interviews

1. Group the children's responses in the interviews into groups which you think demonstrate similar knowledge of the concept being tested. Justify your groupings.

2. What specific implications for the classroom would you draw for these particular Piaget interviews? Explain.

3. Read the other articles about Piaget and then classify each of the students interviewed into one of the stages, sensory-motor, pre-operational, concrete operations, or formal operations. Justify your classification. Do not use age as the justification.

4. What alternatives to the individual interviews employed by Piaget might a teacher employ to assess the stage of intellectual development of his students? Would any of the alternatives you describe provide information comparable to that of a Piagetian interview?

5. Actually conduct some Piagetian interviews with school children. Describe the difficulties you encountered and the results of the interviews. What implications do you see for teaching mathematics to the specific children you interviewed?

6. Design and conduct some interviews with older children (15-18) about the concept of limit. Describe your interviews and attempt to draw conclusions about the children interviewed.

7. Consider some of the perceptual paradoxes commonly given in psychology tests. Taking a Piagetian point of view, what do these paradoxes say about adults?

Mental Growth
and the Art
of Teaching

Irving Adler

How should teachers teach? Before we can answer this question, we must answer first another question: How do pupils learn? The art of teaching, if it is to be effective, must be based on an adequate theory of learning.

Theories of Learning

There have been many different theories of learning, each having different implications for the practice of teaching. These theories may be classified according to their views of the relation between the child and his environment. One type of theory, which saw the environment as the primary active factor, was the basis of the traditional subject-centered school. A second type of theory, which saw the child as the primary active factor, led to the child-centered school. A third type of theory, which recognizes that both the child and his environment play an active role in the learning process, is the foundation for contemporary teaching.

The first type of theory was not always explicitly formulated. But it could easily be recognized as being implicit in school practices which cast the teacher in the role of someone who literally gives the child an education, and cast the child in the role of someone who passively receives it. Various metaphors have been used to describe the theories of this type: the child is a receptacle into which knowledge can be poured; or the child's mind is a *tabula rasa* on which the teacher can write at will; or the child's mind is like clay, to be molded into a socially desirable form.

Theories of this type have been rejected in favor of a different view that underlies the theories of the second and third type. In this view, the child is a self-acting organism that grows. Learning is not a mere accumulation of knowledge, but is a process of growth. This view has led to the emergence of developmental theories of psychology, including some theories of mental growth.

Theories of learning of the first type had pictured the environment as active, and the child as relatively passive. The first critics of these theories, reacting against them, went to the other extreme and formulated theories of the second type, which pictured the child as active, and the environment as relatively passive. Since these theories have influenced contemporary teaching in many ways, both good and bad, it will be worthwhile examining them briefly.

A composite model of some typical theories of the second type would

Irving Adler, "Mental Growth and the Art of Teaching." Reprinted from the *Arithmetic Teacher* (November 1966). ©1966 by the National Council of Teachers of Mathematics. Reprinted with permission.

include propositions like these: Before we can educate the child, we must study the child to find out his nature and needs. The child is growing, and his growth is to be understood as an unfolding from within. The function of the school is to supply growth-promoting nutrients, as soil does to a growing plant. The child learns only through his own activity. This activity should be directed by the child's needs at the moment, and not by any adult purpose. The child learns nothing from words that are divorced from meaning. So verbal learning should be avoided. The growth of the child takes him through several qualitatively distinct stages of development. Instruction during any particular stage should be tailored to what the child needs and is ready for at that stage. If he is not yet ready for a particular type of learning experience, it is necessary to wait until he becomes ready through spontaneous maturation.

When these propositions are subjected to critical analysis and the test of experience, we find that some of them are true and help to guide us toward good teaching practices, while others are false and serve as obstacles to good teaching. It is true that we should study the child we must teach and recognize that he is growing. It is not true that his growth is a process of unfolding from within. An important aspect of his growth is his assimilation of part of the cultural heritage of his generation. His growth is not spontaneous growth. It is directed growth, and the teacher plays an important part in giving it direction. It is true that words divorced from meaning are useless, but it is not true that all verbal learning is useless. In fact, it is chiefly through verbal learning that the child gains access to the knowledge that mankind has accumulated through thousands of years. It is true that we should not teach the child what he is not ready for. But it is not true that we should wait until he is ready. Instead of waiting passively, we should actively help him to become ready.

The principal errors of the theories of the second type are avoided in the recent theories of the third type, which recognize the active role of both the pupil and his environment in the learning process. One of these newer theories is the developmental psychology of Jean Piaget. Since it is based in part on studies of how the child develops his conceptions of number and space, this theory is of particular importance to mathematics teachers and is the subject of the rest of this paper. Piaget's theory of mental growth has many implications for the art of teaching. Let us see what they are.

Piaget's Theory of Mental Growth

Everybody is talking about Piaget, but not many people are reading him. His views are widely misinterpreted and misunderstood. So it is necessary first to give a summary of the principal features of this theory. The summary that I present here lists twelve significant features.

1. Piaget asserts that the basis of all learning is the child's own activity as he interacts with his physical and social environment.

2. The child's mental activity is organized into structures. Separate mental acts are related to each other and grouped together in clusters called "schemas" or patterns of behavior.

3. Mental activity, like metabolic activity, is a process of *adaptation* to the environment. Adaptation consists of two opposed but inseparable

processes, *assimilation* and *accommodation*. Assimilation is the process whereby the child fits every new experience into his pre-existing mental structures. Through the functioning of these structures, he interprets his new experiences in the light of his old experiences. The process of assimilation is a kind of inertia of mental structures, a tendency of these structures to persist. However, the incorporation of new experiences into old structures inevitably modifies them. Accommodation is the process of perpetual modification of mental structures to meet the requirements of each particular experience. Accommodation is the tendency of mental structures to change under the influence of the environment.

4. Mental growth is a social process. The child does not interact with his physical environment as an isolated individual. He interacts with it as part of a social group. Consequently, his social environment mediates between him and his physical environment. The child's interaction with other people plays an important part in the development of his view of the world. It is only through an exchange of ideas with other people that he becomes aware of the one-sided, subjective character of his own point of view. It is only by combining the viewpoints of others with his own that he evolves from having a subjective to having an objective outlook.

5. Although accommodation to the environment leads to a continuous modification of the child's schemas or behavior patterns, the change is not merely quantitative. In the course of time the child's mental structures undergo qualitative changes as well. As the child progresses from infancy to maturity, his characteristic ways of acting and thinking are changed several times as new mental structures emerge out of the old ones, modified by accumulated accommodations. Piaget finds that the child passes through four distinct stages of mental growth, which he calls the *sensori-motor stage,* the *pre-operational stage,* the stage of *concrete operations,* and the stage of *formal operations.*

The child is in the sensori-motor stage from birth to the age of about eighteen months. During this stage, the child's inherited reflexes are gradually modified by experience and combined into complex patterns of behavior. His activity becomes less body centered and more object centered. He becomes aware of the relationship between means and ends in the manipulation of objects. Finally, he becomes aware of the fact that objects exist even when he does not perceive them. Once he has developed the concept of separately existing objects, he is capable of thinking of an object that he does not see. He is then able to plan an action mentally before he carries it out physically. This marks the transition to the next stage of development, the pre-operational stage, in which sensori-motor activity is accompanied by mental activity based on the mental representation of objects and mental anticipations of activities.

The child is in the pre-operational stage from the age of eighteen months to the age of six or seven years. During this stage, the child has already begun to use symbols. They are private symbols at first, and then, as he learns to speak, they are the socially standardized symbols of spoken language. He is egocentric in his view of objects and events. That is, he is incapable of seeing things from any but his own point of view. His thinking is dominated by what he sees at the moment. In his examination of a situation involving several factors that operate at the same time, he tends to fix attention on one factor to the exclusion of the others. In examining a change from one state to another, he fixes his attention on the initial and the final state and ignores the transformation that produced the change. As a result, he is not aware that when two changes occur simultaneously, one may compensate for the other. For example, if a ball of clay is rolled into a sausage shape, its length increases, but its width decreases. The pre-operational child, fixing his attention on the increase in length, will conclude that the sausage has more clay than the original ball because it is longer. He does not see that the decrease in width has compensated for the increase in length. He is not yet aware that the mass of a body is conserved when it is subjected to changes of form or dimension. Similarly, he is not yet aware of the fact that the cardinal number of a set is independent of the arrangement of its members. Since a set of five blocks arranged in a line with spaces between adjacent blocks makes a longer line than the same blocks arranged with no spaces between them, he thinks that the first arrangement has more blocks than the second.

Because of his tendency to fix attention on only one factor at a time, he often comes to contradictory conclusions as he shifts his attention from one factor to another. He will assert the contradictory conclusions without any concern for their inconsistency. Because he tends to neglect transformations from one state to another, he is not aware of the reversibility of many transformations. Because his thinking is dominated by his momentary perception, he tends to associate things with each other by accidents of juxtaposition rather than by any relation of cause and effect or of logical implication. As a result, he has a poor grasp of the relation between part and whole, between element and set, or between subset and set.

The child is in the stage of concrete operations between the ages of seven and eleven. In this stage he has already separated the concept of mass from the concept of length, and the concept of cardinal number from the concept of end-to-end length. He knows that the mass of an object remains unchanged when the arrangement of the set is altered. He has mastered the relationship between set and subset so that he can form hierarchies of sets that are related by inclusion. He has grasped the transitive property of order relations so that he is able to form ordered sets and establish order-preserving one-to-one correspondences.

The words "operations" and "concrete" that Piaget uses to name this stage symbolize the features that distinguish this stage from those that precede and follow it. The operations that Piaget is talking about are mental acts. Piaget defines an *operation* as "an action which can return to its starting point, and which can be integrated with other actions also possessing this feature of reversibility." For example, rolling a ball of clay into a sausage shape is an operation as well as a physical act if the child is aware of it as only one of many changes of form that may be imposed in sequence on the lump of clay, and if he is aware that every such change of form can be reversed or undone. This awareness that actions are reversible and may be combined with other actions makes operational thinking more flexible and more penetrating than pre-operational thinking. Because operations are reversible and can be combined, a system of operations has the structure of a mathematical group, with a law of composition of elements, an identity element, an inverse for every element, and an associative law. The child makes the transition from the pre-operational stage to the stage of concrete operations when his mental acts that used to be isolated and unrelated are finally organized into such grouplike structures.

Piaget calls the mental operations of the child between the ages of seven and eleven "concrete" operations because the starting point for them is always some real system of objects and relations that he perceives. The child in this stage can only organize and order what is immediately present for him. He makes the transition to the next higher stage of thinking ability, the stage of formal operations, when he begins to reason about things that he does not see, when he can reason about the potential as well as the real.

The stage of formal operations, which the child enters near the age of eleven or twelve, is the stage of adult reasoning. In this stage, the child can identify all the possible factors that are relevant to a problem under investigation. He can use combinatorial analysis to form all possible combinations of these factors, one at a time, two at a time, three at a time, and so on. He can formulate hypotheses, draw conclusions from them, and test them against reality. Moreover, he can explore the relations of propositions, not only the relations of things. In short, he is capable of scientific thinking and formal mathematical reasoning. Since formal reasoning involves propositions about propositions, or, in Piaget's terminology, operations on operations, Piaget sometimes refers to formal operations as *operations to the second power.*

6. According to Piaget, perception is not the mere passive registering of raw sensation by the child's sense organs. It is to be understood as *perceptual activity,* in which the child's brain organizes the sensations he gathers in the course of his exploratory activity. His perception of space, for example, is compounded out of images formed on the retina of the eye, the kinesthetic sensations of moving and focusing the eyes, and the coordination of hand and eye in reaching for things, grasping them, moving them, arranging them, and so on. The child's ability to perceive, like his ability to think, undergoes development with age.

7. There is a time lag between the development of a child's ability to perceive a thing and the development of his ability to form a mental image of that thing when it is not perceptually present.

8. In the development of the child's concepts of space, topological notions, such as proximity, separation, order, enclosure, and continuity, arise first. Projective and Euclidean notions arise later.

9. In the development of the child's concept of number, his grasp of order relations and cardinal number grow hand in hand.

10. As the child advances from infancy to maturity, his thinking evolves from a short-range, static, egocentric outlook to a long-range, dynamic coordination of many points of view. This evolution is punctuated by the emergence of a succession of concepts of *invariance:* first the concept of the permanence of the object; then the concepts of the invariance of mass, volume, weight, etc.

11. The order in which a child progresses through the four major stages of mental growth is fixed. But his rate of progress is not fixed. The transition from one stage to the next can be hastened by enriched experience and good teaching. Piaget expresses this idea in these words in his book *The Growth of Logical Thinking from Childhood to Adolescence:*

The maturation of the nervous system can do no more than determine the totality of possibilities and impossibilities at a given stage. A particular social environment remains indispensable for the realization of these possibilities. It follows that their realization can be accelerated or retarded as a function of cultural and educational conditions.

12. Piaget constructed elaborate mathematical models of the mental structures that are characteristic of the stages of concrete operations and formal operations. Since his mathematical models shed little light on the art of teaching, I shall give no details about them.

Misinterpretations of Piaget's Theory

As you can see, even from this sketchy outline, Piaget's theory of mental growth contains many subtleties and complexities. Actually, the theory is so complex that any abbreviated version of it cannot do it justice. Some of the short accounts of the theory that have been prepared for popular consumption eliminate the subtleties and complexities, and thus oversimplify the theory and leave it open to misinterpretation. This is true even of the article that Piaget himself had written for the November 1953 issue of *Scientific American.* Because of the wide currency of these abbreviated accounts of the theory, some gross errors have been made in interpreting and applying it. There are two errors in particular that have had a great influence.

The first of these errors is made by people who interpret Piaget's theory to mean that the stages in a child's mental growth are correlated exclusively with chronological age. On the basis of this interpretation, they conclude that the school cannot hasten the child's progress from one stage to the next, that it must wait for him to mature spontaneously, and that, meanwhile, the best it can do is supply the child with experiences that he is able to assilimate into the mental structure that he has at the moment. Now it is true that Piaget studied only the relationship between mental growth and chronological age. He did not choose to examine the influence of other factors, such as socioeconomic status, methods of teaching, individual differences, and so on. But this does not mean that the other factors have no influence. In fact, Piaget himself has said that there are three factors that play a part in preparing the child to advance from one stage of mental growth to the next. These are "maturation of the nervous system, experience acquired in interaction with the physical environment, and the influence of the social milieu." Moreover, Piaget's theory stresses that while an experience is assimilated into the child's mental structure, the structure is at the same time accommodated to the experience so that it is changed by the experience. The people who fix their attention on only the maturation factor in growth, or on only the assimilation factor in adaptation, and ignore the other factors, are behaving like the child who fixes his attention on only the change in the length of a piece of rolled clay, while he ignores the change in its width. That is, they are in the pre-operational stage in their thinking about mental growth.

The second major error in interpreting Piaget is made by people who emphasize exclusively the child's self-activity in the learning process. These are the enthusiasts for the so-called "discovery" method of teaching, who insist that the child must always be allowed to make his own discoveries without any help from his peers or from his teacher. Now it is true that we should provide every possible opportunity for the child to make, through his own efforts, discoveries that are within his reach. But it is also true that there are many discoveries that are beyond his reach. We cannot expect the child to discover by himself in a few years what all of mankind has discovered in thousands of years. There are many facts and relationships that a child can be led to understand even though he cannot discover them himself. There are some discoveries that he *can* make, if he is guided to them by carefully selected experiences and skillfully formulated questions. Moreover, as Piaget has stressed, the exchange of ideas between a child and his classmates is an important learning experience that should not be sacrificed while we pursue the mirage of each child making all his discoveries by himself in splendid isolation.

If we reject these two errors, it does not mean that we reject any attempt to seek guidance from Piaget's theory for the improvement of the curriculum and methods of teaching. These errors have arisen not from using Piaget's theory but from abusing it. Piaget's theory, properly understood, does have many fruitful implications for the art of teaching. I list a baker's dozen of them in the paragraphs that follow. Because of limitations of time and space, I must describe each of these implications in only a few words. However, exploring all the ramifications of each could easily take several hours of discussion. An adequate treatment of the implications of Piaget's theory of mental growth for the teaching of mathematics would require at least a full semester course.

Implications of Piaget's Theory for the Art of Teaching

Here, then, are some of the implications, as I see them. *162075*

1. Since the child's mental growth advances through qualitatively distinct stages, these stages should be taken into account when we plan the curriculum. We should use two criteria for selecting the mathematical experiences a child should have at any given age: (a) they should be experiences that he is ready for, in view of the stage of mental growth that the child has reached; (b) they should help prepare the child to advance to the next stage. We should not teach a topic too early, but we should also not delay for years a topic that he is ready for. For example, in the traditional mathematics curriculum, the child had little experience with deductive reasoning from hypotheses until he studied geometry at the age of fifteen. However, Piaget's studies show that the child enters the stage of formal operations, the stage of deductive reasoning from hypotheses, at the age of eleven or twelve. It is therefore psychologically sound for us to introduce short units of deductive reasoning from hypotheses as early as the sixth grade. A suitable deductive unit in arithmetic covers the derivation of the rule for multiplying fractions from the associative and commutative laws of multiplication and the rules for multiplying a unit fraction by a whole number or another unit fraction.

2. Before introducing a new concept to the child, test him to be sure that he has mastered all the prerequisites for mastering this concept. If he is not yet ready for the concept, provide him with the experiences that will help him become ready. Of course this is not a new principle. It has been well known to teachers ever since the days of Herbart.

3. The pre-adolescent child makes typical errors of thinking that are characteristic of his stage of mental growth. We should try to understand these errors. Piaget has given us a profile of the thought processes of the pre-operational child and the child in the stage of concrete operations. We should be thoroughly familiar with these profiles. But that is not enough. Besides knowing what errors the child usually makes, we should also try to find out why he makes them. We should keep in mind that an answer or an action that seems illogical from our point of view on the basis of our extensive experience may seem perfectly logical from the child's point of view on the basis of his limited experience. Before we can get the child to understand what we are saying to him, we must first understand in his own terms and on his own ground what the child is saying to us.

For example, when a three-year-old child concludes that a lump of clay rolled into a sausage has more clay than the same lump rolled into a ball, because the sausage is longer than the ball, he is wrong. But in terms of his own limited experience, he is being logical. When he made a "train" with blocks, he observed that by using more and more blocks he could make a longer and longer train. When he ate a banana he observed that the shorter the banana became, the less banana he had left to eat. In short, he had observed a correlation between quantity of matter and length. It is true that the correlation is valid only if other factors, such as width, are kept constant. But his experience had not yet called to his attention the fact that the correlation he had observed had a restricted domain of validity. When he used the correlation to judge the amount of matter in the rolled lump of clay, he was extrapolating the correlation beyond the domain in which its validity had been tested. But it is important to keep in mind that this kind of extrapolation of a rule beyond the domain in which it has been verified is not necessarily a mistake. In fact, scientists are doing it all the time, only they call it scientific prediction. Sometimes the prediction is verified, and we know then that the rule has a broader domain of application. Sometimes the prediction is refuted, and we begin to recognize the limits of the domain in which the rule is valid. Sometimes an impermissible extrapolation leads to the discovery of new knowledge, as the impermissible extrapolation of the square-root operation to negative real numbers led to the discovery of imaginary numbers. The three-year-old is a young scientist who has learned from experience to judge mass by length. With more experience he will discover that this criterion is inadequate.

Actually the three-year-old child does not merely see mass and length as correlated. He goes beyond that and sees them as only barely distinguishable. For the pre-operational child, these two concepts, together with those of volume and cardinal number, have only begun to exist as separate concepts. At first they are all fused into a single vague concept of *amount*. These concepts are fused in the child's mind because they are fused in his experience. The single concept of *amount* becomes differentiated into the separate concepts of mass, volume, length, and cardinal number only as the child's experience establishes the need for this differentiation.

4. We can help the child overcome the errors in his thinking by providing him with experiences that expose them as errors and point the way to the correction of the errors. For example, the pre-operational child judges the number of blocks in a line of blocks by the length of the line,

without regard to the fact that empty spaces as well as blocks may be contributing to the length of the line. To overcome this error, we should have the child perform two pairs of inverse operations: the insertion and removal of spaces, and the insertion and removal of blocks. Many varied experiences of this kind will help to separate the concept of length of the line from the concept of number of blocks in the line. Also, the use of these inverse physical operations will help to crystallize the inverse mental operations that are characteristic of the next stage of growth, the stage of concrete operations.

5. The child in the pre-operational stage tends to fix his attention on one variable to the neglect of others. To help him overcome this error, provide him with many situations like the one just described in which he may explore the influences of two or more variables.

6. A child's thinking is more flexible when it is based on reversible operations. For this reason we should teach pairs of inverse operations in arithmetic together. We should teach that subtraction and addition nullify each other, and that multiplication and division nullify each other.

7. The child in the stage of concrete operations has an incomplete grasp of the relations among the subsets of a set. To close the gaps in his thinking, have him explore by direct observation various sets and their subsets, unions of sets, intersections of sets, and hierarchies of inclusion of sets, as these arise naturally in the learning situation.

8. A prerequisite for the stage of formal operations is the ability to carry out simple combinatorial analysis. All combinatorial analysis is based on the formation of Cartesian products of sets. We can easily teach children systematic ways of forming these products by using tree diagrams and rectangular arrays.

9. Mental growth is encouraged by the experience of seeing things from many different points of view. Although this is especially important for the young child, it should not be neglected in teaching the older child and the adolescent. For example, in the tenth grade we should use many different approaches to the study of geometry. We should use not only the traditional, synthetic approach but also the analytic approach via coordinates, the vector approach, and the approach that is based on the use of isometries of the plane.

10. Physical action is one of the bases of learning. To learn effectively, the child must be a participant in events, not merely a spectator. To develop his concepts of number and space, it is not enough that he look at things. He must also touch things, move them, turn them, put them together, and take them apart. For every new concept that we want the child to acquire, we should start with some relevant action that he can perform. For example, to pave the way for the concept of an angle, we should give him opportunities to turn a hand of a clock or a pointer on a dial.

However, the child's activity should not be kept forever on the level of physical action. The physical action is merely the foundation for the mental operation that we want to develop. We should create opportunities for the child to be less and less dependent on the physical action until the action is entirely internalized as a mental operation. Thus, while we may introduce the addition of integers as a succession of motions on the number line, we should lead the child to discover, as rapidly as possible, ways of doing the addition mentally without recourse to the number line.

11. Since there is a lag between perception and the formation of a mental image, reinforce the developing mental image with frequent use of perceptual data. For example, any time that the child falters in the addition of integers, let him see the addition once more as a succession of motions on the number line.

12. Since mental growth is associated with the discovery of invariants, we should make more frequent use of a systematic search for those features of a situation that remain unchanged under a particular group of transformations. This is both good psychology and good mathematics. In elementary arithmetic, for example, after the equivalence of two sets has been established by a one-to-one correspondence, have the children observe the effect of making substitutions for the elements of the sets or the effect of changing the arrangement of the elements in each set. In geometry, we should certainly make more frequent use of groups of transformations than we do.

13. Piaget pointed out that topological relations are the first geometric relations that are observed by the child, but they are the last ones that were studied explicitly and formally by mathematicians. The reason for this paradox is that topological relations intuitively appear more obvious than either projective or Euclidean relations. Because they are so obvious, the child grasps them early. Also, because they are so obvious, the mathematician tended to take them for granted in his first attempts to develop geometry as an axiomatic structure. There is a lesson for us here concerning the way in which we should deal with topological relations in the schools. We should be careful not to overdo the formalization of the study of topological relations in the tenth grade and below. Children, like the mathematicians of earlier centuries, will not see the need for trying to prove deductively relations that appear to them to be intuitively obvious.

While we take note of the implications of Piaget's theory for the art of teaching, we must also be on guard against possible confusions that may arise from a misreading of his theory. I call attention to two of the possible confusions that may arise from misunderstanding Piaget's use of the terms "concrete operations" and "formal operations."

1. Piaget's use of the word "concrete" in the term "concrete operations" should not be confused with the uses of the word in everyday speech. For Piaget, the concrete operations of a person are mental operations with propositions about some real system of objects and relations that the person perceives. What is concrete or not concrete in this sense is relative to the person's past experience and his mental maturity. To the kindergarten child, uniting a set of two beads with a set of three beads is concrete, but adding the numbers 2 and 3 is not. For the ninth-grade student, the sum $3 + 3$ is concrete, but the sum $x + y$ is not. For the student getting his first introduction to abstract algebra, the additive group of integers is concrete, but the concept of an abstract group is not.

The fact that the stage of mental growth between the ages of seven and eleven is called the stage of concrete operations does not mean that concrete operations are not used at a later age. Concrete operations are used at all stages past the age of seven, but until the age of about eleven they are, in general, the most advanced operations of which the child is capable. Moreover, in the development of new concepts at all stages of learning, it is necessary to proceed from the concrete to the abstract.

2. At the age of eleven or twelve, the child begins to be capable of formal operations, or deductive reasoning from hypotheses. This does not mean that he is incapable of deductive reasoning before then. Concrete operations include deductive reasoning. For example, when the child reasons that $4 + 2 = 4 + 1 + 1 = 5 + 1 = 6$, he is doing deductive reasoning. Opportunities for deductive reasoning should not be neglected in the elementary grades. Of course, the deductive reasoning in this stage should be restricted to reasoning about real objects and relations that the child perceives.

Piaget, through his studies of child development and his theoretical activity, has produced a vast treasury of ideas on how children learn and think. The job of drawing on this treasury for the benefit of the teaching of mathematics has only just begun. I urge that this job be pursued with greater vigor on the basis of a more serious attempt to understand the subtleties and complexities of Piaget's theory.

Exercises: Mental growth and the Art of Teaching

1. Give an example of assimilation. Give an example of accommodation. These two words are "borrowed" from the field of biology; how would a biologist define these two words? Relate these two notions to Jennifer's grappling with the concepts of 1 and 10.

2. Think of your own learning of an advanced abstract mathematical concept. Some people argue that one still goes through the four stages outlined by Piaget even as an adult. Reaction? Justify. Do adults work exclusively at a formal operational level? Give examples to support your answer.

3. Piaget develops his tests in terms of conservation principles. Are there important properties in mathematics that would be better characterized by nonconservation than by conservation? If so, list such properties. If not, list several mathematical principles which reflect the idea of conservation.

4. Are the developmental stages identified by Piaget fixed within the individual or do they vary within the individual relative to the learning task to be accomplished?

5. Suppose you accept the general validity of the Piagetian stages. What does this say to you as a teacher of pre-adolescents? What sort of modifications in curriculum or teaching method might be called for? Give an example to illustrate.

6. Find a mathematical definition for an equivalence relation. For each of the necessary conditions in an equivalence relation, give an example of a corresponding thought process.

7. Examine Adler's implications of the Piagetian research. Do you agree with each of them? Why? Justify. Suppose you did agree, how would this change your behavior with children? Give specific examples.

8. Why are theories of mental growth important to teachers? Why not just wait to teach mathematics until the child's mental growth has been accomplished?

Unit 3

The Nature of
Learning Processes in Mathematics

The structure of mathematics is a guide for determining goals of instruction for most teachers. Each of the authors represented in this section uses structure for more than the analysis of goals. Structure provides a condition to be respected and used in designing and implementing instructional strategies. Each author is committed to a particular psychology and offers specific examples of the applications of these psychologies to instruction. But each author uses and interprets the structure of mathematics differently to make his pedagogical point. Knowledge of the nature of mathematics should complement and strengthen the application of learning processes to instructional decisions.

The logical development of mathematical ideas provides guidance to Gagné in determining appropriate sequences of instructional events. His "Learning and Proficiency" article examines experimental evidence of the nontrivial nature of prior learning in mathematics. The "New Views" article is developed around the same theme of sequencing but identifies critical variables in the application of sequencing strategies. The information processing school of learning theorists represents a new development in the modelling of human behavior.

The structure of particular mathematical situations provides Hartmann with a framework for the application of concepts of Gestalt psychology. Bruner's article appears to be quite consistent with the Gestalt perception of the learning processes. Bruner uses this learner's understanding of the structure of mathematics to supplement the particular problem situations orientation used by Hartmann. Mathematicians of an earlier generation were attracted to the Gestalt explanation of the learning process. Mathematicians of this generation demonstrate the same order of enchantment with Bruner's analyses of learning. What characteristics do these theories share that are so attractive to mathematicians?

The concept of cognitive structure is fundamental to David Ausubel's development of an instructional strategy. The cognitive structure gives the learner a base for anchoring new ideas. The teacher can extend the learner's cognitive structure by arranging today's instruction so that it provides an advance organizer for learning which is yet to come. For example, emphasizing the inverse relationship between the operations of addition and subtraction in grade five provides an advance organizer for equation solving techniques in grade eight.

Zoltan Dienes, like Bruner, uses not only the structure of mathematics but also the nature of doing mathematics as a condition in determining appropriate learning experiences. Each of his primary instructional principles reflects his concern for mathematics being constructive in nature. Learning activities which are essentially passive and receptive are not as conducive to learning effectively as those which help the learner build concepts.

A word of caution is in order: each author uses several words that each of the other authors use. For example, Gagné and Bruner each use the word *concept* as a technical term. The word *structure* is used both in terms of mathematics and as a description of mental processes. As you read, be careful to ask yourself whether each author ascribes the same definition to these common words.

Some New Views of Learning and Instruction

Robert M. Gagné

During recent years there has been an increased recognition of, and even emphasis on, the importance of principles of learning in the design of instruction for the schools. This recognition of the central role of learning in school-centered education seems to be accorded whether one thinks of the instruction as being designed by a teacher, by a textbook writer, or by a group of scholars developing a curriculum.

When the findings of research studies of learning are taken into account, one usually finds questions about instruction to be concerned with such matters as these:

1. For student learning to be most effective, how should the learning task be presented? That is, how should it be communicated to the student?

2. When the student undertakes a learning task, what kinds of activity on his part should be required or encouraged?

3. What provisions must be made to insure that what is learned is remembered and is usable in further learning and problem solving?

Questions such as these are persistent in education. The answers given today are not exactly the same as those given yesterday, and they are likely to be altered again tomorrow. The major reason for these changes is our continually deepening knowledge of human behavior and of the factors which determine it. One should not, I believe, shun such changes nor adopt a point of view which makes difficult the application of new knowledge to the design of novel procedures for instruction. The opportunities for improvement seem great and the risks small.

Status of Learning Research

As a field of endeavor, research on how human beings learn and remember is in a state of great ferment today. Many changes have taken place, and are still taking place, in the conception of what human learning is and how it occurs. Perhaps the most general description that can be made of these changes is that investigators are shifting from what may be called a *connectionist* view of learning to an *information processing* view. From an older view which held that learning is a matter of establishing *connections* between stimuli and responses, we are moving rapidly to acceptance of a view that stimuli are *processed* in quite a number of different ways by the human central nervous system, and that understanding learning is a matter of figuring out how these various processes operate. Connecting one neural event with another may still be

Robert M. Gagné, "Some New Views of Learning and Instruction," *Phi Delta Kappan* (May 1970): 468-472. Reprinted with permission.

the most basic component of these processes, but their varied nature makes connection itself too simple a model for learning and remembering.

My purpose here is to outline some of these changes in the conception of human learning and memory, and to show what implications they may have for the design and practice of instruction. I emphasize that I am not proposing a new theory; I am simply speculating on what seems to me to be the direction in which learning theory is heading.

The Older Conception

The older conception of learning was that it was always basically the same process, whether the learner was learning to say a new word, to tie a shoelace, to multiply fractions, to recount the facts of history, or to solve a problem concerning rotary motion. Edward L. Thorndike held essentially this view. He stated that he had observed people performing learning tasks of varied degrees of complexity and had concluded that learning was invariably subject to the same influences and the same laws.[1] What was this model of learning that was considered to have such broad generalizability?

One prototype is the conditioned response, in which there is a pairing of stimuli, repeated over a series of trials. The two stimuli must be presented together, or nearly together, in time. They are typically associated with an "emotional" response of the human being, such as an eyeblink or a change in the amount of electrical resistance of the skin (the galvanic skin reflex). The size of the conditioned response begins at a low base-line level, and progressively increases as more and more repetitions of the two stimuli are given. Such results have been taken to indicate that repetition brings about an increasingly "strong" learned connection—with an increase in strength that is rapid at first and then more slow.

Learning curves with similar characteristics have been obtained from various other kinds of learned activities, such as simple motor skills like dart-throwing and memorization of lists of words or sets of word-pairs.

Remembering. What about the remembering of such learned activities? Is learning retained better as a result of repetition? Is something that is repeated over and over at the time of learning better recalled after the passage of several weeks or months? The curve which describes forgetting is perhaps equally familiar. Forgetting of such things as lists of nonsense syllables is quite rapid in the beginning, and after several weeks descends to a point at which only about 20 percent is remembered. A motor task is usually retained a great deal better, and after the same amount of time its retention may be as much as 80 percent.

These are the basic facts about remembering. But how is it affected by repetition? Is retention better if the original learning situation has been repeated many times? Evidence is often cited that this is so. Increasing the number of trials of repetition during original learning has the effect of slowing down the "curve of forgetting," i.e., of improving the amount of retention measured at any particular time. Underwood,[2] for example, has stated that "degree of learning" of the task to be recalled is one of the two major factors which influence forgetting in a substantial manner. The second factor is interfering associations, whose strength is also determined by their degree of learning. It should be pointed out that when

Underwood uses the phrase "degree of learning" he refers to amount of practice—in other words, to amount of repetition.

At this point, let me summarize what I believe are the important implications for instruction of what I call the "older" conceptions of learning and memory. The designer of instruction, or the teacher, had to do two major things: First, he had to arrange external conditions of presentation so that the stimulus and response had the proper timing—in other words, so that there was *contiguity* between the presentation of the stimulus and the occurrence of the response. Second, he had to insure that sufficient *repetition* occurred. Such repetition was necessary for two reasons: It would increase the strength of the learned connections; the more the repetition, within limits, the better the learning. Also, repetition was needed to insure remembering—the greater the number of repetitions, the better the retention. Presumably, whole generations of instructional materials and teacher procedures have been influenced in a variety of ways by application of these conceptions of learning to the process of instruction.

Questioning Older Conceptions

During recent years, a number of significant experimental studies of learning and memory have been carried out which call into question some of these older conceptions. (Of course there have always been a certain number of individuals—voices in the wilderness—who doubted that these principles had the general applicability claimed for them.) I shall describe only a few of the crucial new studies here, to illustrate the perennial questions and their possible answers.

Does learning require repetition? A most provocative study on this question was carried out by Rock[3] as long ago as 1957. It has stimulated many other studies since that time, some pointing out its methodological defects, others supporting its conclusions.[4] The finding of interest is that in learning sets of verbal paired associates, practice does not increase the strength of each learned item; each one is either learned or not learned. To be sure, some are learned on the first practice trial, some on the second, some on the third, and so on; but an item once learned is fully learned.

So far as school subjects are concerned, a number of studies have failed to find evidence of the effectiveness of repetition for learning and remembering. This was true in an investigation by Gagné, Mayor, Garstens, and Paradise,[5] in which seventh-graders were learning about the addition of integers. One group of children was given four or five times as many practice problems on each of 10 subordinate skills as were given to another group, and no difference appeared in their final performance. A further test of this question was made in a study of Jeanne Gibson,[6] who set out to teach third- and fourth-graders to read decimals from a number line. First, she made sure that subordinate skills (reading a number in decimal form, writing a number in decimal form, locating a decimal number on a number line) were learned thoroughly by each child. One group of students was then given a total of 10 practice examples for each subordinate skill, a second group 25 for each, and a third none at all. The study thus contrasted the effects of no repetition of learned skills, an intermediate amount of repetition, and a large amount of repetition. This variable was not found to have an effect on performance, both when tested immediately after learning and five weeks later. Those students who practiced repeated examples were not shown to do better, or to remember better, than those who practiced not at all.

Still another study of fairly recent origin is by Reynolds and Glaser,[7] who used an instructional program to teach 10 topics in biology. They inserted frames containing half as many repetitions, in one case, and one-and-a-half times as many repetitions, in another, as those in a standard program. The repetitions involved definitions of technical terms. When retention of these terms was measured after an interval of three weeks, the investigators were unable to find any difference in recall related to the amount of repetition.

I must insert a caveat here. All of the studies I have mentioned are concerned with the effects of repetition immediately after learning. They do not, however, test the effect of repetition in the form of *spaced reviews*. Other evidence suggests the importance of such reviews; in fact, this kind of treatment was found to exert a significant effect in the Reynolds and Glaser study. Note, though, that this result may have quite a different explanation than that of "strengthening learned connections."

Modern Conceptions of Learning

Many modern learning theorists seem to have come to the conclusion that conceiving learning as a matter of strenthening connections is entirely too simple. Modern conceptions of learning tend to be highly analytical about the events that take place in learning, both *outside* the learner and also *inside*. The modern point of view about learning tends to view it as a complex of processes taking place in the learner's nervous system. This view is often called an "information-processing" conception.

One example of an information processing theory is that of Atkinson and Shiffrin.[8] According to this theory, information is first registered by the senses and remains in an essentially unaltered form for a short period of time. It then enters what is called the short-term store, where it can be retained for 30 seconds or so. This short-term store has a limited capacity, so that new information coming into it simply pushes aside what may already be stored there. But an important process takes place in this short-term memory, according to Atkinson and Shiffrin. There is a kind of internal reviewing mechanism (a "rehearsal buffer") which organizes and rehearses the material even within this short period of time. Then it is ready to be transferred to long-term store. But when this happens it is first subjected to a process called *coding*. In other words, it is not transferred in raw form, but is transformed in some way which will make it easier to remember at a later time. Still another process is *retrieval*, which comes into play at the time the individual attempts to remember what he has learned.

It is easy to see that a much more sophisticated theory of learning and memory is implied here. It goes far beyond the notion of gradually increasing the strength of a single connection.

Prerequisites for learning. If repetition or practice is not the major factor in learning, what is? The answer I am inclined to give is that the most dependable condition for the insurance of learning is the prior learning of prerequisite capabilities. Some people would call these "specific readinesses" for learning; others would call them "enabling conditions." If one wants to insure that a student can learn some specific new activity, the very best guarantee is to be sure he has previously learned the prerequisite capabilities. When this in fact has been accomplished, it seems to me quite likely that he will learn the new skill without repetition.

Let me illustrate this point by reference to a study carried out by Virginia Wiegand.[9] She attempted to identify all the prerequisite capabilities needed for sixth-grade students to learn to formulate a general expression relating the variables in an inclined plane. Without using the exact terminology of physics, let us note that the task was to formulate an expression relating the *height* of the plane, the *weight* of the body traversing downwards, and the *amount of push* imparted to an object at the end of the plane. (Wiegand was not trying to teach physics, but to see if the children could learn to formulate a physical relationship which was quite novel to them.) The expression aimed for was, "Distance pushed times a constant block weight equals height of plane times weight of cart."

Initially, what was wanted was explained carefully to the students; the plane and the cart were demonstrated. Thirty students (out of 31) were found who could not accomplish the task; that is, they did not know how to solve the problem. What was it they didn't know? According to the hypothesis being investigated, they didn't know some *prerequisite* things. Figure 1 shows what these missing intellectual skills were thought to be.

What Wiegand did was to find out which of these prerequisite skills were present in each student and which were not present. She did this by starting at the top of the hierarchy and working downwards, testing at each point whether the student could do the designated task or not. In some students, only two or three skills were missing; in others, seven or eight. When she had worked down to the point where these subordinate capabilities *were* present, Wiegand turned around and went the other way. She now made sure that all the prerequisite skills were present, right up to, but not including, the final inclined plane problem.

The question being asked in this study was, If all the prerequisite skills are present, can the students now solve this physical problem which they were unable to solve previously? Wiegand's results are quite clear-cut. Having learned the prerequisites, nine out of 10 students were able to solve the problem which they were initially unable to solve. They now solved the problem without hesitation and with no practice on the problem itself. On the other hand, for students who did not have a chance to learn the prerequisites, only three of 10 solved the problem (and these were students who had no "missing" skills). This is the kind of evidence that makes me emphasize the critical importance of prerequisite intellectual skills. Any particular learning is not at all difficult if one is truly prepared for it.

Coding and remembering. Quite a number of studies appear in the experimental literature pertaining to the effects of coding of information

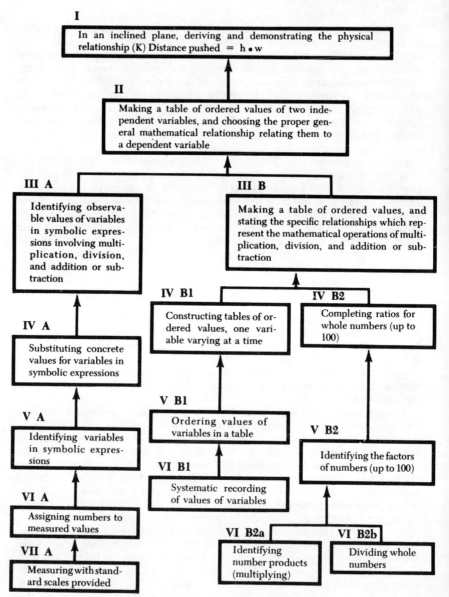

Figure 1.

A hierarchy of subordinate intellectual skills applicable to the problem of deriving a general expression relating variables in an inclined plane (Wiegand, 1969).

on its retention. I choose as an illustration a study by Bower and Clark.[10] These investigators studied the recall by college students of 12 lists of 10 nouns apiece. In learning each list, each student was encouraged to make up a story connecting the nouns. For each student there was a yoked control who was not encouraged to make up a story but who was permitted the same amount of time to learn each list of nouns.

Here is an example of a story which one of the subjects constructed for the words *vegetable, instrument, college, nail, fence, basin, merchant, queen, scale,* and *goat:* "A vegetable can be a useful instrument for a college student. A carrot can be a nail for your fence or basin. But a merchant of the queen would scale that fence and feed the carrot to a goat."

The subjects were asked to recall each list immediately after their study of it. They recalled 99 percent under both conditions. The subjects were later asked to recall all of the lists, after they had learned all 12. In this case there was an enormous difference: The recall of the narrative group averaged 93 percent, that of the non-narrative group only 13 percent. In other words, deliberate coding had increased recall by seven times.

Retrieval and remembering. Suppose that learning has indeed occurred; what will insure that whatever has been learned will be remembered? There seems to be at least some absence of evidence that simply practicing or repeating things after they have been learned has the effect of improving retention. What the individual does when he is asked to remember something is to *retrieve* it; that is, he brings to bear a process of searching and finding, in his memory, something he is looking for. This process is probably very little understood at present, but there is increasing evidence that it does occur and that it plays a crucial role in remembering.

Some interesting work has been done on the subject of retrieval. In one experiment, Tulving and Pearlstone[11] had groups of high school students learn lists of words of various lengths: 12 words, 24 words, or 48 words. The words themselves were instances of categories,, such as four-footed animals (cow, rat); weapons (bomb, cannon); forms of entertainment (radio; music); professions (lawyer, engineer), and so on. The words were presented one at a time in mixed-up order. Different lists also used one, two, or four words in each category.

Once the lists of words had been learned, recall was measured under two different conditions. In the first, the learners were simply told to write down all the words they could remember. In the second, the category names were used as cues to recall; that is, the learners were asked to write down all the words they remembered which were "forms of entertainment," all which were "four-footed animals," and so on. These extra cues worked wonders on recall. The effect was more marked the greater the number of words that had to be recalled. The differences among those learning 48 words was striking, amounting to a twofold increase.

These results show in a rather clear way how powerful is the effect of such extra cues on retrieval of information that has been learned. In this study, the words themselves can be said to have been "equally well learned" in all the groups. What was different between the groups was the aid they were given in retrieving what they had learned. This is only one of the accumulating pieces of evidence that remembering is markedly affected by retrieval at the time of recall, more than it is, perhaps, by events taking place at the time of learning.

Implications for Instruction

The contrasts between older and newer conceptions of learning and memory seem to me quite remarkable. What implications do they have for instruction? If there are indeed newly discovered ways to affect learning and remembering, how might they be put to use in the classroom and in materials of the curriculum?

First, there is the very fundamental point that each learner approaches each new learning task with a different collection of previously learned prerequisite skills. To be effective, therefore, a learning program for each child must take fully into account what he knows how to do already, and what he doesn't know how to do already. One must find out what prerequisites he has already mastered—not in a general sense, but in a very precise sense for each learner. Does this mean one must do "diagnostic testing"? Yes, that's exactly what it means. To do so, of course, one must first develop the requisite diagnostic tests. By and large, we don't have them.

Second, the most important guide to the learning that needs to be accomplished is the set of prerequisites that the student has not yet mastered. Remember here Wiegand's experiment. When she systematically saw to it that students climbed the hierarchy, skill by skill, this was what was specifically needed to get them to engage in the problem solving they were originally unable to do.

Third, do students need additional practice to insure retention? If by this is meant, "Should they be given many additional examples so that what they have learned will be 'strengthened'?," I think the evidence says it probably won't work this way. Periodic and spaced reviews, however, are another matter, and it seems likely that these have an important role to play in retention. Notice that when a review is given the student has to exercise his strategies of retrieval.

This brings me to the final point, which concerns the processes of coding and retrieval. Probably what should be aimed for here is the learning by students of strategies of coding. These are by no means the same as what are called "mnemonic systems," although it is possible that such systems have a contribution to make in teaching us how coding might be done. For meaningful learning, it appears even more likely that notions like "advance organizers" and "anchoring ideas," as studied by Ausubel,[12] may be particularly powerful.

Similarly, retrieval strategies are also a class of objective that might be valued for instruction. From the evidence we have, I should say that retrieval strategies might very well consist in networks of superordinate categories into which newly learned specific information, or specific intellectual skills, can be placed. Having students learn to retrieve information by a process of search which first locates such superordinate networks may be a major way of providing them with the capability of good retention.

Even these two or three aspects of modern learning conceptions, it seems to me, lead to a very different view of what instruction is all about. In the most general sense, instruction becomes not primarily a matter of communicating something that is to be stored. Instead, it is a matter of stimulating the use of capabilities the learner already has at his disposal, and of making sure he has the requisite capabilities for the present learning task, as well as for many more to come.

References

1. E. L. Thorndike, *Human Learning*. New York: Appleton-Century, 1931, p. 160.

2. B. J. Underwood, "Laboratory Studies of Verbal Learning," in E. R. Hilgard (ed.), *Theories of Learning and Instruction. Sixty-third Yearbook, Part I.* Chicago: National Society for the Study of Education, 1964, p. 148.

3. I. Rock, "The Role of Repetition in Associative Learning," *American Journal of Psychology,* June, 1957, pp. 186-93.

4. W. K. Estes, B. L. Hopkins, and E. J. Crothers, "All-or-None and Conservation Effects in the Learning and Retention of Paired Associates," *Journal of Experimental Psychology,* December, 1960, pp. 329-39.

5. R. M. Gagné, J. R. Mayor, H. L. Garstens, and N. E. Paradise, "Factors in Acquiring Knowledge of a Mathematical Task," *Psychological Monographs,* No. 7, 1962 (Whole No. 526).

6. J. R. Gibson, "Transfer Effects of Practice Variety in Principle Learning." Berkeley: University of California. Ph.D. Dissertation, 1964.

7. J. H. Reynolds and R. Glaser, "Effects of Repetition and Spaced Review upon Retention of a Complex Learning Task," *Journal of Educational Psychology,* October, 1964, pp. 297-308.

8. R. C. Atkinson and R. M. Shiffrin, "Human Memory: A Proposed System and Its Control Processes," in K. W. Spence and J. T. Spence (eds.), *The Psychology of Learning and Motivation: Advances in Research and Theory,* Vol. 2. New York: Academic Press, 1968, pp. 89-195.

9. V. K. Wiegand, "A Study of Subordinate Skills in Science Problem Solving." Berkeley: University of California. Ph.D. Dissertation, 1969.

10. G. H. Bower and M. C. Clark, "Narrative Stories as Mediators for Serial Learning," *Psychonomic Science,* April, 1969, pp. 181-82.

11. E. Tulving and Z. Pearlstone, "Availability Versus Accessibility of Information in Memory for Words," *Journal of Verbal Learning and Verbal Behavior,* August, 1966, pp. 381-91.

12. D. P. Ausubel, *Educational Psychology: A Cognitive View.* New York: Holt, Rinehart and Winston, 1968.

**Exercises: Some New Views
of Learning and Instruction**

1. Cite two or three examples from classroom mathematics where the "older" conceptions of learning and memory are applied today. Show the emphasis on contiguity and repetition.

2. Gagné argues that the "modern" conception of learning is an "information-processing" conception. He uses words like *short-term storage, limited capacity, rehearsal buffer, coding,* and *retrieval.* These are all computer words. Are we now guilty of thinking that the human mind is just like a computer? What evidence can you give from children learning mathematics that would justify this approach? What evidence can you give that would not justify this approach?

3. What explanations can be offered to explain why the subjects in the study by Bower and Clark who made up stories were better able to remember the lists later?

4. If one accepts the importance of the role of retrieval in remembering, what implications would this have for the teaching of a specific mathematics topic of your choice? Give your example and explain.

5. How does the information-processing conception of learning imply the importance of prerequisites for learning? Is Gagné axe-grinding here? Explain.

6. If practice is not effective in "strengthening" what is learned, what role does practice play when it is spaced?

7. Why aren't "memory systems" the same as strategies of coding? For example,

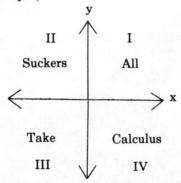

can be used to remember which trigonometric functions are positive in each quadrant. Is this a good strategy of coding?

Gestalt Psychology and Mathematical Insight

George W. Hartmann

This paper is presented as an expression of belief and hope that some of the newer formulations in psychological theory will bring about a democratization of mathematical competence somewhat akin to the astonishing elevations of performance often obtained in the field of reading. The symbolism and thought processes involved in using conventional language are not essentially different from those employed in the more universal terminology and operations of mathematics. If reading ability is now better and more widespread among Americans than at any previous time in our history, there is no reason why a better comprehension of numbers, figures, and conceptual relations should not make quantitative inferences as common as simple words and sentences. Although it is probably true that algebra and geometry *as ordinarily taught* are unable to become more influential because of the present intellectual and motivational limitations of the average learner, I should like to maintain the position that by taking more adequate advantage of the principles of mental development, our teachers could make the special modes of thought of the mathematician a part of the daily routine of our citizenry. The deserved success of the Bell and Hogben volumes reveals what some of these possibilities are.

Most persons will probably approve of this end and accept the claim just made, but many maintain that the existing crisis in secondary-school mathematics — a situation which affects more than the jobs of certain teachers of specialized subject-matter — is a consequence of altered social conditions and changed educational attitudes, and that a better methodology and modernized psychology can do little more than make possible a graceful retreat to a more humble role. The essential problem, these folks say, is a curricular one. True enough; but it would be a mistake to assume that there are no psychological foundations to a curriculum, mathematical or otherwise. The human values in both content and procedure are too intimately allied to permit that. It is a common error which holds that an educational discipline becomes "progressive," i.e., modern and enlightened, by skillfully and effectively teaching that which should not be taught and not teaching that which should. The *How* and the *Why* of instruction are organically related and a truly satisfactory solution for one will tend simultaneously to solve the other.

Insofar as any distinct psychological system has been adopted by mathematics teachers, their training appears to have led them to favor

Reprinted from the October 1937 issue of *The Mathematics Teacher* (XXX, 265-70). © 1966 by the National Council of Teachers of Mathematics. Used by permission.

the connectionist or "bond" theory of learning, although the frank hostility of this position to the claims of the formal disciplinarians has led many of them to cling desperately to the long-discredited notion of separate mental "faculties." This occurred, one may suppose, not because mathematicians are any more conservative than other pedagogues, but because such an outlook supported their claims to a preferred status in the conventional course of study. In the last decade, however, a small but growing group has found a more satisfactory foundation for its practice in the tenets of the Gestalt brand of psychology — a theory to which mathematicians are perhaps temperamentally congenial because it literally outlines a subtle "geometry of the mind." What are some of the considerations upon which this advanced (and advancing) viewpoint rests?

In my judgment, there are three propositions which are basic to that type of theorizing which goes by the name of Gestalt:

1. *All experience or mental life implies a differentiation of the sensory or perceptual field to which the organism can respond into some kind of figure-ground pattern.* In other words, there must be heterogeneity of stimulation before any psychological process can occur. If we were affected by nothing but undifferentiated homogeneous energy, e.g., a single flat level of grey in vision or an unvarying tonal mass in hearing, the very conditions for consciousness itself would probably be absent. Difference produces phenomena. From this standpoint, variety is more than the spice of life — it is a prerequisite of life itself.

This dualism of figure and ground is an inescapable feature of all perception, but it is a *functional* and not a structural antithesis. The figure is simply that feature of the situation to which primary attention is given *at the moment* — the ground, although essential to the emergence of the figure, has a secondary role in terms of the focalization of the organism's interest. In Figure 26 (which is typical of all "reversible"

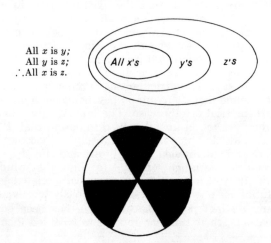

Figure 26. **Ground-Figure Relationship.**

Note how the black and white propellers are alternately seen.

patterns), the black and the white areas alternate in dominating the reader's field; when the black region is figure, the white is ground, and vice versa. Most of the patterns we encounter are far more stable than this, although all can be reorganized subjectively and made to fluctuate to some extent. Thus, the interlinear white space on this page, which is normally "unnoticed" ground even though it is absolutely essential to the reading act, can — with some effort — undergo a transformation and acquire temporarily the status of "figure."

2. *The course of mental development is from a broad, vague, and indefinite total to the particular and precise detail.* The end-result of this process of differentiation is an organized body of "clear and distinct ideas" — that state of affairs so dear to the mind of the skilled logician. But it is far from the condition with which the growth process starts. Sharpness of outline is what we end with, not what we have at the beginning. The act of perceiving is normally initiated by a dim general awareness of the object; it is only as this continues to act upon the observer that its internal "structure" emerges.

In the light of this conception many mathematical commonplaces acquire a new meaning. Euclidean geometry, e.g., is a masterpiece of reasoning, but it would be far truer to the facts of genetic psychology if the order in which it is commonly presented were completely reversed. Its logic is atomistic or elementaristic,[1] i.e., it begins with the most

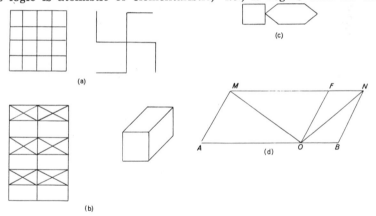

Figure 27. Figure and Outline Perception.

How "field forces" govern what is discriminated. Can you isolate the swastika and the "box" in their companion figures on the left?

---

[1] The following clipping from the humorous column of a teachers' journal indicates that many persons have become aware of the artificial and "absurd atomism" implied in many of the textbook problems of an older day.

"Arithmetic is a science of truth," said the professor earnestly. "Figures can't lie. For instance, if one man can build a house in 12 days, 12 men can build it in one."

"Yes," interrupted a quick-brained student, "Then 288 will build it in one hour, 17,280 in one minute, and 1,036,800 in one second. And I don't believe they could lay one brick in that time."

highly refined and mature abstractions, such as "definitions" (note the term with its suggestion of optical focusing!) of *point* and line, rather than starting with more massive and "natural" percepts like cubes, surfaces, etc. Psychological experimentation indicates that a "point" is a fairly late and high-grade achievement of one's spatial understanding. Our visual-tactile world is not originally made up of *points* — instead, these emerge from it. Paradoxical as it may seem in the light of the usual placement of courses, solid or tri-dimensional geometry is the source of all later spatial analysis which ends in, but does not proceed from, the strange entity that "has" location without extension. It has even been argued that division is a more primitive arithmetical operation than addition.

3. *The properties of parts are functions of the whole or total system in which they are imbedded.* In perception this principle is clearly observed by the fact that a grey square upon a blue ground looks yellowish and the *same* grey patch on a yellow field appears bluish (color induction or "contrast"). In Figure 27(c), most persons "see" a square and a hexagon in contact (presumably because the structural organization of the drawing favors this response), but the capital letter K, which is just as much present in substance, is usually not discerned. To isolate a familiar portion of the alphabet in this situation requires the segregation of two markedly dependent parts of two strongly unified "figures." Frequency and repetition cannot account for this phenomenon, for even the most experienced geometer has encountered a K more often than he has seen these elementary patterns in contact. The external arrangement as given compels a corresponding internal organization.

Figure 27(d) is even more impressive because it exhibits some of the mechanism underlying an illusion. Most persons, if asked to compare the diagonals MO and ON, will unhesitatingly declare MO the longer. Actually the two are drawn of equal length. The effect is apparently traceable to the larger rectangle $MAOF$ which causes the observer to "see" (not to "infer" in the usual logical sense) its diagonal as greater than that of the noticeably smaller rectangle $FOBN$. The "illusion" is partly overcome by erecting a perpendicular at O, thus minimizing the unanalyzed and unequal influence of the two major areas. A better example of the Gestaltist's claim that a line is functionally a derivative of a plane could hardly be found.

It seems probable that the meaning of a number in series is a special case of membership-character being conditioned by its role in some structure. Thus, the number "364" is comparatively meaningless in isolation. Conceptually, however, its fuller meaning is necessarily derived from some schema, such as "less than 400," "between 350 and 400," "nearer 400 than 300," etc. In the case of lightning calculators, most numbers have acquired some such "individuality" as this — a fact which contributes something to an understanding of their ability.

With these three generalizations as one's conceptual tools, it is surprising how many obscure phenomena swiftly become more intelligible. A number of years ago, while examining the arithmetical errors of college students, I noticed that a mistake was more likely to occur when a larger number was being added to a smaller one than in the converse case. Thus, $9+7$ and $7+9$ both make 16 but addition errors are decidedly more frequent with the latter combination. The rule has also been statistically verified for two-place numbers, and I am inclined to believe it holds for fractions and any other number combinations. If we view the addition of two quantities as a simple case of completing an indicated total, then this observation is brought under the head of "closure" phenomena. In any language completion test, gaps are to be filled in, and the test's difficulty is roughly proportional to the number and extent of the gaps involved. This "totalizing effect" is seen in the figures below. A "circle" with three-fourths of circumference visible is easily seen as a full circle at a slight distance from the eye — a fact occasionally used by the oculist in visual testing. An arc equivalent to a quarter-circle does not lend itself so readily to the "restoration" of the entire circle (cf. Figure 28). *Pari passu,* when one adds 9 to 7, one is traversing a greater psychic distance than when one adds 7 to 9. Hence, the goal, "16," is more surely reached in the latter instance. Since there is no inherent difference in the difficulty of the two symbolic number combinations, the variation in accuracy must be traced to the

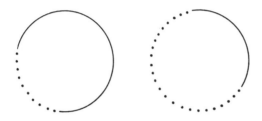

Figure 28. Closure Effects.

*Closure effects are easier, less ambiguous, and more impres-
sive the less the perceptual "filling" required.*

forces behind the schematic structure by which they are represented in the organism.

If this explanation makes concepts obey the same laws as percepts, that has its source in the conviction that thinking and reasoning are dependent upon the processes of perceiving. The latter activity is nearer to the concrete situation than the former, and ordinarily involves less of the inferential closure represented by the dotted portions of Figure 28. But in the case of "Euler's circles" as used in elementary demonstrations of formal logic, one literally "sees" how intimately syllogistic proof is linked to direct sensory perception of the basic pattern. It seems that the famous Swiss mathematician of the eighteenth century was once a tutor by correspondence to a dull-witted Russian princess and devised this method of convincing her of the reality and necessity of certain relations established deductively. Thus, the syllogism can be tested for its truth or falsity if diagrammed as indicated. Here the concentric "ellipses" reproduce the essentials of the situation so faithfully that the answer becomes a matter of "mere inspection."

This process of making an organism aware of the conditions governing the phenomena to which it is reacting is essentially what is meant by the "insight" experience. Rightly construed, insight is not a peculiarity of the "higher" rational functions, but a process necessarily occurring at all mental levels. Simple adjustments of bodily position, such as lifting the foot to go up a step or bending the knees when accepting an invitation to sit down, constantly exhibit a low but decidedly real level of insight. In every case there is a rearrangement of an action pattern as a result of changing forces in the "field." If the new organization fits the needs of the situation, the neural stresses and strains return to a state of relative dynamic equilibrium and the "problem" is said to be "solved."

A simple geometrical example will suffice to clarify this all-important point. Given Figure 29(a), a high school sophomore is asked to find the area of a circumscribed square, knowing only the radius of the inscribed circle. A question or assignment such as this normally produces

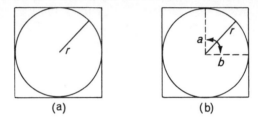

(a)   (b)

Figure 29. Quantitative Relations.

How insight operates with quantitative relations.

a mild tension which does not vanish until an answer satisfactory to the organism is achieved. If one examines carefully what occurs, it is plain that a state of mild bewilderment sets in which lasts until a configuration emerges. Figure 29(b) is typical of a wide range of educational situations. The pupil is blocked as long as he sees the key item (the radius r) *fixed* in its first position; but as soon as he *shifts* it mentally to position a or b, its properties are *transformed* — it is no longer a half-diameter, but is now half of a side of the square. This reorganization accomplished, the response $A = (2r)^2$ is immediately and confidently made. The delay involved is almost entirely taken up by the time required to bring Gestalt principle No. 3 (above) into action.[2]

Another basic conception of Gestalt theory which promises to be useful in mathematical instruction is the idea of transposition. This had its origin in the common observation that a musical melody is not the same as the sum of the separate tones that presumably comprise it. "The Star Spangled Banner" can be sung by bass, soprano, and other voices, or played by various orchestral instruments in different "keys." In none of these instances need any of the individual tones be alike, and

---

[2]Alternative and equally simple solutions are also possible.—On one occasion a mathematics teacher, who was listening to an *oral* exposition of this example, insisted the answer was wrong because he had heard and interpreted $(2r)^2$ as $2(r)^2$. It is significant that erroneous as well as correct solutions are equally intelligible under this principle.

yet the common pattern or melodic sequence is easily recognized. The melody is the whole which is transposed and *transposable* from one situation to another.

Is it not conceivable that most of the "abstractions" with which mathematics deals are of this nature? The ratio between the diameter of any circle and its circumference is expressed by the constant π, and this relation is transposable whether the actual concrete materials that constitute the circle are made of copper, rope or carbon and liquid particles. In the case of the equation — which is plainly the heart of most mathematical operations — we see how the procedure is built around the possibility of preserving "permanence amid change." One may think of the equation as a step-wise expanding or contracting pattern that preserves invariant the condition of equality or identity throughout the various transformations that may be legitimately applied to it.

This situation was long ago appreciated by gifted thinkers and accounted for much of the esteem in which mathematical method was held. However, it would be a great mistake to assume that other aspects of nature and behavior are lacking in this respect.[3] Even a goldfish (who is known to be less intelligent than the humble cockroach) reacts appropriately to a group of "equivalent stimuli." Thus, suppose he has learned that food is to be found behind that light which is intermediate in intensity among three sources of illumination in his field. Now change the absolute brightness of the three bulbs by doubling or halving the "intensity" of the current. In either event, the goldfish swims unhesitatingly toward that light which is relatively in the "middle" of the triad. This response — found widely in all species — is difficult or impossible to explain on the basis that learning is specific, i.e., restricted in its effectiveness to the precise stimulus-object employed. This interpretation itself must be false as it is killed by its own hyper-specificity. No two learning situations can be alike in *all* respects; the likeness is found in the organization of the wholes and not in the substance of the pieces. A wooden table is functionally more like a metal table than it is like a wooden chair. The way things are put together determines the attributes of the system thus established.

It must be obvious from these illustrations that mathematical and psychological research have more in common than is usually realized. The field and organismic approaches to behavior have more than a purely physical and biological connotation, and point set theory is heavily used in systems of "topological" psychology. Much of this is the inevitable and desirable consequence of the unity of scientific thought. The contributions of mathematical technique to psychological advance have been most impressive, but the reverse type of obligation has rarely been incurred, presumably because there was so little to borrow! Perhaps if mathematics teachers act upon the recognition that the content of their discipline has to be rediscovered and created *de novo* by every learner, they will have provided themselves with the one tool needed to make the Grand Tradition of rigorous thinking influential in the lives of our people.

---

[3]Consequently geometry cannot be the only area in which rigorous proof is possible. No proof can rise above the assumptions and hidden theorems upon which it rests.

Exercises: Gestalt Psychology and Mathematical Insight

1. Consider Hartman's analogy of the U. S. population's reading ability and its ability in mathematics. Is it possible to dramatically increase the general population's ability in mathematics? Hartmann wrote this article in 1937; have we now achieved general mathematical literacy? Explain. Give examples.

2. Give a mathematical example from the classroom of the Gestalt figure-ground aspect of learning. Identify the figure and the ground and describe the effect of varying each and the relationship of each to mathematics instruction.

3. Argue, by citing examples from mathematics, that "mental development is from a broad, vague and indefinite total to the particular and precise detail." Where, in the Cronbach article of Unit 1, can you find support for this position.

4. Give a mathematical example to illustrate that "the properties of parts are functions of the whole or total system in which they are embedded." Suggest implications for teaching.

5. How would you relate Piaget to Gestalt psychology? Examples? Assimilation? Accommodation?

On Learning Mathematics

Jerome S. Bruner

I am challenged and honored to be asked to speak before a group of teachers of mathematics on the nature of learning and thinking—particularly mathematical learning and thinking. Let me introduce you to my intentions by citing a remark of the English philosopher, Weldon. He noted that one could discriminate between difficulties, puzzles, and problems. A difficulty is a trouble with minimum definition. It is a state in which we know that we want to get from here to there, both points defined rather rawly, and with not much of an idea how to bridge the gap. A puzzle, on the other hand, is a game in which there is a set of givens and a set of procedural constraints, all precisely stated. A puzzle also requires that we get from here to there, and there is at least one admissible route by which we can do so, but the choice of route is governed by definite rules that must not be violated. A typical puzzle is that of the Three Cannibals and Three Missionaries, in which you must get three missionaries and three cannibals across a river in a boat that carries no more than two passengers. You can never have more cannibals than missionaries on one side at a time. Only one cannibal can row; all three missionaries can. Another puzzle, one in which the terminus has not yet been achieved, is the so-called Twin Primes Conjecture. Now, Weldon proposes that a problem is a difficulty upon which we attempt to impose a puzzle form. A young man, trying to win the favor of a young lady—a difficulty—decides to try out successively and with benefit of correction by experience, a strategy of flattery—an iterative procedure, and a classic puzzle—and thus converts his difficulty into a problem. I rather expect that most young men do all this deciding at the unconscious level; I hope so for the sake of my daughters! But the point of mentioning it is not my fatherly jealousy, but to emphasize that the conversion of difficulties into problems by the imposition of puzzle forms is often not always done with cool awareness, and that part of the task of the mathematician is to work toward an increase in such awareness. But this gets me ahead of my exposition.

Let me urge that the pure mathematician is above all a close student of puzzle forms—puzzles involving the ordering of sets of elements in a manner to fulfill specifications. The puzzles, once grasped, are obvious, so obvious that it is astounding that anybody has difficulty with mathematics at all, as Bertrand Russell once said in exasperation. "Why, the rowing cannibal takes over another cannibal and returns. Then he takes over the other cannibal and returns. Then two missionaries go over, and one of them brings back a nonrowing cannibal. Then a missionary

---

Jerome S. Bruner, "On Learning Mathematics," *The Mathematics Teacher* (December 1960): 610-619. Reprinted with permission of the author.

takes the rowing cannibal over and brings back a nonrowing cannibal. Then two missionaries go over and stay, while the rowing cannibal travels back and forth, bringing the remaining cannibals over one at a time. And there are never more cannibals than missionaries on either side of the river." It is simple. If you say that my statement of the solution is clumsy and lacking in generality, even though correct, you are quite right. But now we are talking mathematics.

For the mathematician's job is not pure puzzle-mongering. It is to find the deepest properties of puzzles so that he may recognize that a particular puzzle is an exemplar—trivial, degenerate, or important, as the case may be—of a family of puzzles. He is also a student of the kinship that exists between families of puzzles. So, for example, he sets forth such structural ideas as the commutative, associative, and distributive laws to show the manner in which a whole set of seemingly diverse problems all have a common puzzle form imposed on them.

It is probably the case that there are two ways in which one goes about both learning mathematics and teaching it. One of them is through a technique that I want to call unmasking: discovering the abstracted ordering properties that lie behind certain empirical problem solutions in the manner in which the triangulation techniques used for reconstructing land boundaries in the Nile valley eventually developed into the abstractions of plane geometry, having first been more like surveying than mathematics. Applied mathematics, I would think, is still somewhat similar in spirit, although I do not wish to become embroiled in the prideful conflict over the distinction between pure and applied. The more usual way in which one learns and teaches is to work directly on the nature of puzzles themselves—on mathematics *per se*.

I should like to devote my discussion to four topics related to the teaching or learning of mathematics. The first has to do with the role of *discovery,* wherein it is important or not that the learner discover things for himself. I have been both puzzled (or I should say "difficultied") and intrigued hearing some of you discuss this interesting matter. The second topic is *intuition,* the class of nonrigorous ways by which mathematicians speed toward solutions or cul-de-sacs. The third is mathematics as an analytic language, and I should like to concentrate on the problem of the *translation* of intuitive ideas into mathematics. I hope you will permit me to assume that anything that can be said in mathematical form can also be said in ordinary language, though it may take a tediously long time to say it and there will always be the danger of imprecision of expression. The fourth and final problem is the matter of *readiness:* when is a child "ready" for geometry or topology or a discussion of truth tables? I shall try to argue that readiness is factorable into several more familiar issues.

Discovery

I think it can be said now, after a decade of experimentation, that any average teacher of mathematics can do much to aid his or her pupils to the discovery of mathematical ideas for themselves. Probably we do violence to the subtlety of such technique by labelling it simply the "method of discovery," for it is certainly more than one method, and each teacher has his own tricks and approach to stimulating discovery by the student. These may include the use of a Socratic method, the devising of particularly apt computation problems that permit a student to find regularities, the act of stimulating the student to short cuts by which he discovers for himself certain interesting algorisms, even the projection of an attitude of interest, daring, and excitement. Indeed, I am struck by the fact that certain ideas in teaching mathematics that take a student away from the banal manipulation of natural numbers have the effect of freshing his eye to the possibility of discovery. I interpret such trends as the use of set theory in the early grades partly in this light—so too the Cuisenaire rods, the use of modular arithmetic, and other comparable devices.

I know it is difficult to say when a child has discovered something for himself. How big a leap must he take before we will grant that a discovery has been made? Perhaps it is a vain pursuit to try to define a discovery in terms of what has been discovered by whom. Which is more of a discovery—that $3 + 4 = 7$, that $3x + 4x = 7x$, or that 7 shares with certain other sets the feature that it cannot be arranged in rectangular ranks? Let me propose instead that discovery is better defined not as a product discovered but as a process of working, and that the so-called method of discovery has as its principal virtue the encouragement of such a process of working or, if I may use the term, such an attitude. I must digress for a moment to describe what I mean by an attitude of discovery, and then I shall return to the question of why such an attitude may be desirable not only in mathematics but as an approach to learning generally.

In studying problem solving in children between the ages of 11 and 14, we have been struck by two approaches that are almost polar opposites. Partly as an analogy, but only partly, we have likened them, respectively, to the approach of a listener and the approach of a speaker toward language. There are several interesting differences between the two. The listener's approach is to take the information he receives in the order in which it comes; he is bound in the context of the flow of speech he is receiving, and his effort is to discern a pattern in what comes to him. Perforce, he lags a bit behind the front edge of the message, trying to put the elements of a moment ago together with those that are coming up right now. The listener is forced into a somewhat passive position since he does not have control of the direction of the message or of its terminus. It is interesting that listeners sometimes fall asleep. It is rare for a speaker to fall asleep. For the speaker is far more active. He, rather than lagging behind the front edge of the message he is emitting, is well out ahead of it so that the words he is speaking lag behind his thoughts. He decides upon sequence and organization.

Now a wise expositor knows that to be effective in holding his auditor he must share some of his role with him, must give him a part in the construction game by avoiding monologue and adopting an interrogative mode when possible. If he does not, the listener either becomes bored or goes off on his own internal speaking tour.

Some children approach problems as a listener, expecting to find an answer or at least some message there. At their best they are receptive, intelligent, orderly, and notably empirical in approach. Others approach problem solving as a speaker. They wish to determine the order of information received and the terminus of their activity and to march ahead of the events they are observing. It is not only children. As a friend of mine put it, a very perceptive psychologist indeed, some men are more interested in their own ideas, others are more interested in nature. The fortunate ones care about the fit between the two. Piaget, for example, speaks of the two processes of accommodation and assimilation, the former being a process of accepting what is presented and changing with it, the latter being the act of converting what one encounters into the already existing categories of one's thought. Each attitude has its excesses. The approach of the listener can become passive and without direction. The approach of the speaker can become assimilative to the point of autistic thinking. As Piaget points out in his brilliant studies of thinking in early childhood, some sort of balance between the two is essential for effective cognitive functioning.

It is in the interest of maintaining this balance that, I would propose, the approach of discovery is centrally important. The overly passive approach to learning, the attitude of the listener, creates a situation in which the person expects order to come from outside, to be in the material that is presented. Mathematical manipulation requires reordering, unmasking, simplification, and other activities akin to the activity of a speaker.

There is one thing that I would emphasize about discovery: its relation to reward and punishment. I have observed a fair amount of teaching in the classroom: not much, but enough to know that a great deal of the daily activity of the student is not rewarding in its own right. He has few opportunities to carry a cycle of working or thinking to a conclusion, so that he may feel a sense of mastery or of a job well done. At least when he makes a paper airplane, he can complete the cycle almost immediately and know whether or not the thing flies. It is not surprising then that it is necessary to introduce a series of extrinsic rewards and punishments into school activity—competition, gold stars, etc.—and that, in spite of these, there are still problems of discipline and inattention. Discovery, with the understanding and mastery it implies, becomes its own reward, a reward that is intrinsic to the activity of working. I have observed and even taught classes in which the object was to stimulate discovery, and I have seen masterful teachers accomplish it. I am impressed by the fact that, although competitive advantage is still strong in such a classroom atmosphere, it is nonetheless the case that the experience of discovering something, even if it be a simple short cut in computation, puts reward into the child's own hands.

I need not tell you that there are practical difficulties. One cannot wait forever for discovery. One cannot leave the curriculum entirely open and let discovery flourish willy-nilly wherever it may occur. What kinds of discoveries to encourage? Some students are troubled and left out and have a sense of failure. These are important questions, but they should be treated as technical and not as substantive ones. If emphasis upon discovery has the effect of producing a more active approach to learning and thinking, the technical problems are worth the trouble.

Intuition

It is particularly when I see a child going through the mechanical process of manipulating numbers without any intuitive sense of what it is all about that I recall the lines of Lewis Carroll: "Reeling and Writhing, of course, to begin with . . . and then the different branches of Arithmetic—Ambition, Distraction, Uglification, and Derision." Or as Max Beberman puts it, much more gently, "Somewhat related to the notion of discovery in teaching is our insistence that the student become aware of a concept before a name has been assigned to the concept."[1] I am quite aware that the issue of intuitive understanding is a very live one among teachers of mathematics and even a casual reading of the Twenty-fourth Yearbook[2] of your Council makes it clear that you are also very mindful of the gap that exists between proclaiming the importance of such understanding and actually producing it in the classroom.

Intuition implies the act of grasping the meaning or significance or structure of a problem without explicit reliance on the analytic apparatus of one's craft. It is the intuitive mode that yields hypotheses quickly, that produces interesting combinations of ideas before their worth is known. It precedes proof; indeed, it is what the techniques of analysis and proof are designed to test and check. It is founded on a kind of combinatorial playfulness that is only possible when the consequences of error are not overpowering or sinful. Above all, it is a form of activity that depends upon confidence in the worthwhileness of the process of mathematical activity rather than upon the importance of right answers at all times.

I should like to examine briefly what intuition might be from a psychological point of view and to consider what we can possibly do about stimulating it among our students. Perhaps the first thing that can be said about intuition when applied to mathematics is that it involves the embodiment or concretization of an idea, not yet stated, in the form of some sort of operation or example. I watched a ten-year-old playing with snail shells he had gathered, putting them into rectangular arrays. He discovered that there were certain quantities that could not be put into such a rectangular compass, that however arranged there was always "one left out." This of course intrigued him. He also found that two such odd-man-out arrays put together produced an array that was rectangular, that "the left out ones could make a new corner." I am not sure it is fair to say this child was learning a lot about prime numbers. But he most certainly was gaining the intuitive sense that would make it possible for him later to grasp what a prime number is and, indeed, what is the structure of a multiplication table.

I am inclined to think of mental development as involving the construction of a model of the world in the child's head, an internalized set of structures for representing the world around us. These structures are organized in terms of perfectly definite grammars or rules of their own, and in the course of development the structures change and the grammar that governs them also changes in certain systematic ways. The way in which we gain lead time for anticipating what will happen next and what to do about it is to spin our internal models just a bit faster than the world goes.

Now the child whose behavior I was just describing had a model of quantities and order that was implicitly governed by all sorts of

---

[1]Max Beberman, *An Emerging Program of Secondary School Mathematics* (Cambridge, Massachusetts: Harvard University Press, 1958), p. 33.

[2]*The Growth of Mathematical Ideas, Grades K-12,* Twenty-fourth Yearbook of the National Council of Teachers of Mathematics (Washington, D. C.: The National Council, 1959).

seemingly subtle mathematical principles, many of them newly acquired and some of them rather strikingly original. He may not have been able to talk about them, but he was able to do all sorts of things on the basis of them. For example, he had "mastered" the very interesting idea of conservation of quantity across transformations in arrangement or, as you would say, the associative law. Thus, the quantity 6 can be stated as $2 + 2 + 2$, $3 + 3$, and by various "irregular" arrangements, as $2 + 4$, $4 + 2$, $2 + (3 + 1)$, $(2 + 3) + 1$, etc. Inherent in what he was doing was the concept of reversibility, as Piaget calls it, the idea of an operation and its inverse. The child was able to put two sets together and to take them apart; by putting together two prime number arrays, he discovers that they are no longer prime (using our terms now) but can be made so again by separation. He was also capable of mapping one set uniquely on another, as in the construction of two identical sets, etc. This is a formidable amount of highbrow mathematics.

Now what do we do with this rather bright child when he gets to school? Well, in our own way we communicate to him that mathematics is a logical discipline and that it has certain rules, and we often proceed to teach him algorisms that make it seem that what he is doing in arithmetic has no bearing on the way in which one would proceed by nonrigorous means. I am not, mind you, objecting to "social arithmetic" with its interest rates and baseball averages. I am objecting to something far worse, the premature use of the language of mathematics, its end-product formalism, that makes it seem that mathematics is something new rather than something the child already knows. It is forcing the child into the inverse plight of the character in *Le Bourgeois Gentilhomme* who comes to the blazing insight that he has been speaking prose all his life. By interposing formalism, we prevent the child from realizing that he has been thinking mathematics all along. What we do, in essence, is to remove his confidence in his ability to perform the processes of mathematics. At our worst, we offer formal proof (which is necessary for checking) in place of direct intuition. It is good that a student know how to check the conjecture that Sx is equivalent to the expression $3x + 5x$ by such a rigorous statement as the following: "By the commutative principle for multiplication, for every x, $3x + 5x = x3 + x5$. By the distributive principle, for every x, $x3 + x5 = x(3 + 5)$. Again by the commutative principle, for every x, $x(3 + 5) = (3 + 5)x$ or Sx. So, for every x, $3x + 5x = Sx$." But it is hopeless if the student gets the idea that this and this only is *really* arithmetic or algebra or "math" and that other ways of proceeding are really for nonmathematical slobs. Therefore, "mathematics is not for me."

I would suggest, then, that it is important to allow the child to use his natural and intuitive ways of thinking, indeed to encourage him to do so, and to honor him when he does well. I cannot believe that he has to be taught to do so. Rather, we would do well to end our habit of inhibiting the expression of intuitive thinking and then to provide means for helping the child to improve in it. To this subject I turn next.

Translation

David Page wrote me last year: "When I tell mathematicians that fourth grade students can go a long way into 'set theory,' a few of them reply, 'Of course.' Most of them are startled. The latter ones are completely wrong in assuming that set theory is intrinsically difficult. Of course, it may be that nothing is intrinsically difficult—we just have to wait the centuries until the proper point of view and corresponding language is revealed!" How can we state things in such a way that ideas can be understood and converted into mathematical expression?

It seems to me there are three problems here. Let me label them the *problem of structure,* the *problem of sequence,* and the *problem of embodiment.* When we try to get a child to understand a concept, leaving aside now the question of whether he can "say" it, the first and most important problem, obviously, is that we as expositors understand it ourselves. I apologize for making such a banal point, but I must do so, for I think that its implications are not well understood. To understand something well is to sense wherein it is simple, wherein it is an instance of a simpler, general case. I know that there are instances in the development of knowledge in which this may not prove to be the case, as in physics before Mendeleev's table or in contemporary physics where particle theory is for the moment seemingly moving toward divergence rather than convergence of principles. In the main, however, to understand something is to sense the simpler structure that underlies a range of instances, and this is notably true in mathematics.

In seeking to transmit our understanding of such structure to another person—be he a student or someone else—there is the problem of finding the language and ideas that the other person would be able to use if he were attempting to explain the same thing. If we are lucky, it may turn out that the language we would use would be within the grasp of the person we are teaching. This is not, alas, always the case. We may then be faced with the problem of finding a homologue that will contain our own idea moderately well and get it across to the auditor without too much loss of precision, or at least in a form that will permit us to communicate further at a later time.

Let me provide an example. We wish to get across to the first-grade student that much of what we speak of as knowledge in science is indirect, that we talk about such things as pressure or chemical bonds or neural inhibition although we never encounter them directly. They are inferences we draw from certain regularities in our observations. This is all very familiar to us. It is an idea with a simple structure but with complicated implications. To a young student who is used to thinking of things that either exist or do not exist, it is hard to tell the truth in answer to his question of whether pressure "really" exists. We wish to transmit the idea that there are observables that have regularities and constructs that are used for conserving and representing these regularities, that both, in different senses, "exist," and the constructs are not fantasies like gremlins or fairies. That is the structure.

Now there is a sequence. How do we get the child to progress from his present two-value logic of things that exist and things that do not exist to a more subtle grasp of the matter? Take an example from the work of Inhelder and Piaget. They find that there are necessary sequences or steps in the mastery of a concept. In order for a child to understand the idea of serial ordering, he must first have a firm grasp on the idea of comparison—that one thing includes another or is larger than another. Or, in order for a child to grasp the idea that the angle of incidence is equal to the angle of reflection, he must first grasp the idea that for any angle at which a ball approaches a wall, there is a corresponding unique angle by which it departs. Until he grasps this idea, there is no point in talking about the two angles being equal or bearing any particular relationship to each other, just as it is a waste to try to explain transitivity to a child who does not yet have a firm grasp on serial ordering.

The problem of embodiment then arises: how to embody illustratively the middle possibility of something that does not quite exist as a clear and observable datum? Well, one group of chemists working on a new curriculum proposed as a transitional step in the sequence that the child be given a taped box containing an unidentified object. He may do anything he likes to the box: shake it, run wires through it, boil it, anything but open it. What does he make of it? I have no idea whether this gadget will indeed get the child to the point where he can then more easily make the distinction between constructs and data. But the attempt is illustrative and interesting. It is a nice illustration of how one seeks to translate a concept (in this instance the chemical bond) into a simpler homologue, an invisible object whose existence depended upon indirect information, by the use of an embodiment. From there one can go on.

The discussion leads me immediately to two practical points about teaching and curriculum design. The first has to do with the sequence of a curriculum, the second with gadgetry. I noted with pleasure in the introductory essay of the Twenty-fourth Yearbook of the National Council of Teachers of Mathematics that great emphasis was placed upon continuity of understanding: "Theorem 2. Teachers in all grades should view their task in the light of the idea that the understanding of mathematics is a continuum. . . . This theorem implies immediately the corollaries that: (1) Teachers should find what ideas have been presented earlier and deliberately use them as much as possible for the teaching of new ideas. (2) Teachers should look to the future and teach some concepts and understandings even if complete mastery cannot be expected." Alas, it has been a rarity to find such a structure in the curriculum, although the situation is likely to be remedied in a much shorter time than might have been expected through the work of such organizations as the School Mathematics Study Group. More frequently fragments are found here and there: a brilliant idea about teaching co-ordinate systems and graphing, or what not. I have had occasion to look at the list of teaching projects submitted to the National Science Foundation. There is everything from a demonstrational wind tunnel to little Van de Graaff generators, virtually all divorced from any sequence. Our impulse is toward gadgetry. The need instead is for something approximating a spiral curriculum, in which ideas are presented in homologue form, returned to later with more precision and power, and further developed and expanded until, in the end, the student has a sense of mastery over at least some body of knowledge.

There is one part of the picture in the building of mathematical curriculum now in progress where I see a virtual blank. It has to do with the investigation of the language and concepts that children of various ages use in attempting intuitively to grasp different concepts and sequences in mathematics. This is the language into which mathematics will have to be translated while the child is en route to more precise mastery. The psychologist can help in all this, it seems to me, as a handmaiden to the curriculum builder, by devising ways of bridging the gap between ideas in mathematics and the students' ways of understanding such ideas. His rewards will be rich, for he not only will be helping education toward greater effectiveness, but also will be learning afresh about learning. If I have said little to you today about the formal psychology of learning as it now exists in many of our university centers, it is because most of what exists has little bearing on the complex and ordered learning that you deal with in your teaching.

Readiness

One of the conclusions of the Woods Hole Conference of the National Academy of Sciences on curriculum in science was that any subject can be taught to anybody at any age in some form that is honest.[3] It is a brave assertion, and the evidence on the whole is all on its side. At least there is no evidence to contradict it. I hope that what I have had to say about intuition and translation is also in support of the proposition.

Readiness, I would argue, is a function not so much of matura-tion—which is not to say that maturation is not important—but rather of our intentions and our skill at translation of ideas into the language and concepts of the age we are teaching. But let it be clear to us that our intentions must be plain before we can start deciding what can be taught to children of what age, for life is short and art is long and there is much art yet to be created in the transmission of knowledge. So let me say a word about our intentions as educators.

When one sits down to the task of trying to write a textbook or to prepare a lesson plan, it soon becomes apparent—at whatever level one is teaching—that there is an antinomy between two ideals: coverage and depth. perhaps this is less of a problem in mathematics than in the field of history or literature, but not by any means is it negligible. In content, positive knowledge is increasing at a rate that, from the point of view of what portion of it one man can know in his life-time, is, to some, alarming. But at the same time that knowledge increases in its amount, the degree to which it is structured also increases. In Robert Oppenheimer's picturesque phrase, it appears that we live in a "multi-bonded pluriverse" in which, if everything is not related to everything else, at least everything is related to something. The only possible way in which individual knowledge can keep proportional pace with the surge of available knowledge is through a grasp of the relatedness of knowledge. We may well ask of any item of information that is taught or that we lead a child to discover for himself whether it is worth knowing. I can only think of two good criteria and one middling one for deciding such an issue: whether the knowledge gives a sense of delight and whether it bestows the gift of intellectual travel beyond the information given, in the sense of containing within it the basis of generalization. The middling criterion is whether the knowledge is useful. It turns out, on the whole, as Charles Sanders Peirce commented, that useful knowledge looks after itself. So I would urge that we as school men let it do so and concentrate on the first two criteria. Delight and travel, then.

It seems to me that the implications of this conclusion are that we opt for depth and continuity in our teaching rather than coverage, and that we re-examine afresh what it is that bestows a sense of intellectual

---

[3]Jerome S. Bruner, *The Process of Education* (Cambridge, Massachusetts: Harvard University Press, 1960).

delight upon a person who is learning. To do the first of these, we must ask what it is that we wish the man in our times to know, what sort of minimum. What do we mean by an educated man? There is obviously not time now to examine this question in the detail it deserves. But I think we would all agree that, at the very least, an educated man should have a sense of what knowledge is like in some field of inquiry, to know it in its connectedness and with a feeling for how the knowledge is gained. An educated man must not be dazzled by the myth that advanced knowledge is the result of wizardry. The way to battle this myth is in the direct experience of the learner—to give him the experience of going from a primitive and weak grasp of some subject to a stage in which he has a more refined and powerful grasp of it. I do not mean that each man should be carried to the frontiers of knowledge, but I do mean that it is possible to take him far enough so that he himself can see how far he has come and by what means.

If I may take a simple example, let me use the principles of conservation in physics: the conservation of energy, mass, and momentum. Indeed, I would add to the list the idea of invariance across transformation in order to include mathematics more directly. The child is told, by virtue of living in our particular society and speaking our particular language, that he must not waste his energy, fritter it away. In common experience, things disappear, get lost. Bodies "lose" their heat; objects set in motion do not appear to stay in motion as in the pure case of Newton's law. Yet, the most powerful laws of physics and chemistry are based on the conception of conservation. Only the meanest of purists would argue against the effort to teach the conservation principles to a first-grade student on the grounds that it would be "distorted" in the transmission. We know from the work of Piaget and others that, indeed, the child does not easily agree with notions based on conservation. A six-year-old child will often doubt that there is the same amount of fluid in a tall, thin glass jar as there was in a flat, wide one, even though he has seen the fluid poured from the latter into the former. Yet, with time and with the proper embodiment of the idea—as in the film of the Physical Science Study Committee where a power plant is used as an example—the idea can be presented in its simplest and weakest form.

Let the idea be revisited constantly. It is central to the structure of the sciences of nature. In good time, many things can be derived from it that yield tremendous predictive power. Coverage in this sense, that is, showing the range of things that can be related to this particular and powerful something, serves the ends of depth. But what of delight? If you should ask me as a student of the thought processes what produces the most fundamental form of pleasure in man's intellectual life, I think I would reply that it is the reduction of surprise and complexity to predictability and simplicity. Indeed, it is when a person has confidence in his ability to bring off this feat that he comes to enjoy surprise, to enjoy the process of imposing puzzle forms upon difficulties in order to convert them into problems. I think we as educators recognized this idea in our doctrine of the "central subject," the idea of co-ordinating a year's work around a central theme. But choosing a central theme horizontally, for the year's work, is arbitrary and often artificial. The central themes are longitudinal. The most important central theme is growth in your own sense of mastery, of knowing today that you have more power and control and mastery over a subject than you had last year. If we produce such a sense of growth, I think it produces delight in knowledge as a by-product automatically.

My choice of the conservation theorems as an illustration was not adventitious. I tried to choose one as basic to the natural sciences as one could make it. Similar themes recur and have eventual crescendo value in other fields: the idea of biological continuity whereby giraffes have giraffe babies and not elephant babies, and idea of tragedy in literature, the notion of the unit of measure in mathematics, the idea of chance as a fraction of certainty in statistics, the grammar of truth tables in logic. It would seem to be altogether appropriate to bring about a joining of forces of experienced teachers, our most gifted scholars, and psychologists to see what can be done to structure longitudinal curricula of this order.

When we are clear about what we want to do in this kind of teaching, I feel reasonably sure that we will be able to make rapid strides ahead in dealing with the pseudoproblem of readiness. I urge that we use the unfolding of readiness to our advantage: to give the child a sense of his own growth and his own capacity to leap ahead in mastery. The problem of translating concepts to this or that age level can be solved, the evidence shows, once we decide what it is we want to translate.

I have perhaps sounded optimistic in my remarks. The evidence warrants optimism, and I cannot help but feel that we are on the threshold of a renaissance in education in America. Let me recapitulate my argument briefly. With the active attitude that an emphasis on discovery can stimulate, with greater emphasis (or fewer restraints) on intuition in our students, and with a courteous and ingenious effort to translate organizing ideas into the available thought forms of our students, we are in a position to construct curricula that have continuity and depth and that carry their own reward in giving a sense of increasing mastery over powerful ideas and concepts that are worth knowing, not because they are interesting in a trivial sense but because they give the ultimate delight of making the world more predictable and less complex. It is this perspective that makes me optimistic and leads me to believe that our present flurry is the beginning not of another fad, but of an educational renaissance.

Exercises: On Learning Mathematics

1. In Bruner's discussion of discovery he seems to be talking about two ways of learning, one he identifies with Piaget's accommodation and the other he identifies with Piaget's assimilation. Give two examples from classroom mathematics which illustrate the differences between these two ways of learning. Is Bruner's distinction a valuable one? Why? What implications does Bruner's discussion of discovery have for your teaching? Argue for your position.

2. What point is Bruner trying to make about intuition? List arguments for and against Bruner's position. To behave as Bruner suggests, what would you do in your classroom?

3. At the beginning of the readiness section Bruner quotes his now famous remark that "any subject can be taught to anybody at any age in some form that is honest." How can this remark possibly be justified? What arguments does Bruner give to support his statement? Take the notion of derivative. How would you deal with the fundamental idea of derivative with first graders? Seventh graders? Outline specific activities. (You may notice that this activity forces you to learn something new about derivatives yourself.)

4. Bruner states that ideas must be raised constantly, that central themes are longitudinal. This is often referred to as the "spiral approach." What are the central themes in mathematics? Choose one and indicate how it could fit into the mathematics curriculum at grades two, eight, eleven, fifteen (junior in college).

5. What relationship do you see between Bruner's views and the views of Gagné? Explain.

6. Give what you see as the practical implications of the points made by Bruner in this article for your own teaching.

A Theory of Mathematics-Learning

Z. P. Dienes

Before going into any theory of mathematics-learning, I ought first of all to make quite clear what is meant by mathematics in this connection. It is not to be thought of as an elaboration of a number of techniques, although such techniques are clearly essential for the effective use of mathematics. Mathematics will be regarded rather as a structure of relationships, the formal symbolism being merely a way of communicating parts of the structure from one person to another. A mathematical statement is a statement of some connection within the structure; to express such a connection we make use of a symbolism, which is in effect a kind of language invented purely for this purpose. For example, the symbolic statement

$$2(A + B) = 2A + 2B$$

states a connection between two parts of the structure, the part dealing with addition and the part dealing with multiplication. The knowledge that we can pass from the symbols $2(A + B)$ to the symbols $2A + 2B$ and vice versa is technical knowledge, which *may not include* any knowledge of the actual connecting link symbolized in the formula. I have already shown that such formal statements about structures are continually being made in our schools without the structures themselves being understood.

By mathematics I shall therefore understand *actual* structural relationships between concepts connected with numbers (pure mathematics), together with their applications to problems arising in the real world (applied mathematics). The learning of mathematics I shall take to mean the apprehension of such relationships together with their symbolization, and the acquisition of the ability to apply the resulting concepts to real situations occurring in the world.

It is difficult to see how any 'stimulus-response' theory of learning can be applied to learning defined in this way. Such theories regard learning as a process of conditioning certain responses which can subsequently be evoked by certain stimuli. But if we look again at the *average* mathematics lesson, it is just such a conditioning process that we shall see at work. Stimuli are presented, which are linked to certain responses called 'correct answers' by some form of explanation (this being the only part of the process paying any attention to the structure). A reward system, in some cases reinforced by a punishment system, conditions subjects to give the 'correct answers'. In the majority of cases, as long as

Z. P. Dienes, "A Theory of Mathematics-Learning," *Building Up Mathematics* 3rd ed., chapter 2 (London: Hutchinson Publishing Group Ltd.), 1967. Reprinted with permission.

the responses continue to be the 'correct answers' no further action is taken. The accent on the whole is on 'getting it right', i.e. on establishing a certain specified response to a specified stimulus; and reference to the structure is only made as an aid to bringing about this state of affairs. The pupils getting habitually 'wrong answers' are usually the ones whose understanding has not kept pace with the growth of the structure. They are reduced to learning certain tricks in order to increase the number of 'correct answers' which they feel compelled to give in the conditioning situation. It is by no means certain that even those who get mostly 'correct answers' really understand the parts of the structure to which these answers refer. What is it about mathematics which makes a stimulus-response learning-situation less adequate than in other subjects? It is that the accent in mathematics is more on structure and less on content. In the teaching of history the most important thing about historical events is that they happened, although at a more advanced stage attempts are made to 'structure' events by considering them as parts of a pattern. This kind of patterning is, on the other hand, the very essence of mathematical thinking. Moreover, established patterns soon come to be regarded as mathematical objects, which are then fitted into further patterns; these in turn, upon becoming familiar, are regarded as objects, and so on. This superimposed patterning or structuring goes on with alarming rapidity for the mathematically uninitiated, who very soon find themselves left behind in the race.

It will be worth our while to examine this process a little more closely. To make the matter as clear as possible I shall use the more familiar grammatical terminology of 'subjects' and 'predicates', instead of the logical language of 'elements' and 'classes'. Let us take the concept of natural number. The predicate 'there are three' refers to a collection of things. This collection of things is the subject of the above predicate, i.e. of the predicate stating that there are three things in the collection. Going beyond this stage we might, upon seeing three apples and three oranges, be tempted to say: 'there are as many apples as there are oranges'. We are now applying a new predicate, 'there being as many as'. What is the subject of this predicate? Certainly not the collection. The subject is the number of things in one of the collections. This number, the number three, was being used just now as a predicate applying to a collection. It is now being used as a subject, to which another predicate is applied, the predicate 'there being as many as'. To be precise, this predicate has really two subjects, the number of things in the apple collection and the number of things in the orange collection. What we are saying is that these numbers are the same. This would be even clearer if we considered the predicate 'there being less than'. If there are two oranges and three apples, we can apply the predicate 'there being less than' to the oranges and the apples and say 'there are less oranges than there are apples'. The subjects are the numbers 2 and 3, and the predicate is 'is less than', which is written symbolically

$$2 < 3$$

In the beginning 2 and 3 were saying something about collections of things, then the predicates 'is the same as' or 'is less than' were saying something about 2 and 3. It is easy to go on and say things about 'is the same as'. For example, to make clear what is meant by adding, the concept 'is the same as' is required. When we count up to 2, and then we count 3 more, we find that we get to the 'same' number as if we count straight up to 5. Similarly, multiplication needs addition before it can be explained, and the formula 2(A + B) needs both addition and multiplication; that is, this formula says something about an enormous number of things, each of which says things about other things, and so on, until we get back to numbers, which say something about collections of things. Putting it grammatically, predicates become subjects for further predicates, which in turn become subjects for yet further predicates, and the sky is the limit in this mathematical race.

People who are good at taming predicates and reducing them to a state of subjection are good mathematicians. Each time a mathematician creates a predicate, he almost immediately begins to wonder what he can say about the new predicate. The establishment of a predicate applying to a certain class of subjects is a kind of enclosure round them: a mathematically minded person will soon get mathematical claustrophobia in such enclosed spaces and begin to wonder what is outside the fence—in other words, he will begin to look for connections between his predicates. You cannot keep him enclosed for long. If this kind of endless open thinking is the essence of mathematical thinking, then it is clearly radically different from the more mundane, everyday type of thinking. Psychologists who have studied the problems of learning and thinking have seldom been mathematicians; perhaps this is the reason why no adequate theory has so far been proposed to account for the kind of learning that takes place in this rather special field.

There is, however, a small kernel of experimental evidence collected by mathematically oriented research workers in psychology, and on this may be based a feasible skeleton theory of mathematics-learning. Much, however, remains to be done. If the problems are put before the teaching profession, teachers in the field will be able to join actively in collecting data and thus help to build a solid foundation for a theory.

The sources from which our skeleton theory will derive are the well-known researches of Piaget, the work of the Cognition Project at Harvard led by Bruner, the fascinating work done by Sir Frederick Bartlett, and certain experimental results of my own. The reader is referred to the bibliography for detailed references.

A few words about these researches will be necessary. Piaget was the first to see that the process of forming a concept takes far longer than had been believed, and that much work, seemingly unrelated to the concept, must be done before there is any clue to the direction which the thinking is taking. This is the largely unconscious or play stage, where the ingredients of the concept are played with long before there is any idea that these ingredients will one day help to classify events in the world in a useful way. The baby plays with sounds and syllables long before he has any idea that later these sounds will be vehicles of communication. The child plays with bricks or other objects, grouping them in collections of different shapes and sizes, long before he knows that he is really practising the ingredients for later number and spatial concepts. We have experiences of fluctuations in prices and incomes, long before we try to coordinate these experiences according to any economic theory. Clearly the concepts in each case could never be formed without extensive play with their ingredients. The second stage is ushered in by the slow

realization of a direction along which our experiences can gradually be
built into a meaningful whole. The baby begins to realize that certain
sounds occur whenever certain events occur, as, for example, that his
sister always appears when her name is pronounced. He gradually
begins to try to produce the sounds in *appropriate* circumstances; he is in
fact consciously trying to move in the direction of meaningful
communications. The child playing with bricks eventually realizes that
those collections which contain two objects have something in common:
for example, that there is one for him and one for mother. This is the dawn
of mathematical experience, experience which will lead to its climax,
much later, in the apprehension of pure number. In our reflections on
prices and salaries, a time may come when we feel we should try to
understand the relationships involved, and we may go to the library and
borrow a book on economics. This second stage sooner or later leads on to
the third stage, when somehow the picture clicks into focus and we feel we
'understand'. The closing of the cycle thus comes with the sudden
realization of an end-point in a mental journey. (9) This is followed by a
period of practice, in order presumably to anchor the new concept into our
experience more firmly and so make it part of our operational armoury in
dealing with the perplexities of our environment. The baby that has
learnt to say 'Mummy' will go on saying it over and over again, just to see
whether it has the supposed effect. The child who has discovered number
will probably go on quite interminably building a number of identical
towers and the like, and moreover insist that adults share in this
repetition although they very quickly tire of the monotony. To
understand the child's point of view we need merely to remember how we
ourselves tend to bore our friends with any new-found theory, and try to
apply it in the most unsuitable circumstances! This is the practice stage
following the realization of a concept, which in turn will act as the play
stage for the next crop of concepts. So the cycles go on, one after the other,
each one building on previously performed cycles.

The reader will find no difficulty in seeing that this dynamic
description of the learning process is more likely to fit the facts of
mathematical learning than any atomistic, stimulus-response description. But of course it is only a framework, and the framework must
be filled in with the content of what it is that we learn. Situations differ
from one another, for example in logical structure. We may have to relate
sets of experiences by means of different logical connections. We may
have to learn that events A and B always happen together, as for
instance the ringing of a bell and the starting of a lesson. We associate the
two events by *conjoining* them into a joint event; we make a conjunction
of two previously unconnected events. In another situation, we know that
if there are two people on a short list for a post, only one of them can be
appointed to the post. We unite these events by *disjoining* the two
separate possibilities; we make a disjunction of two events which were
unconnected before the short list was drawn up. In the conjunctive case
we say:

'The bell rings *and* the lesson begins.'

In the disjunctive case we say:

'*Either* Mr. Brown gets the post *or* Mr. Green gets it.'

There are a number of other logical connections which we make between already established concepts. For example: *If* Mr. A is a Londoner, *then* I can speak to him in English. It is not true that if I can speak to someone in English I am necessarily in the presence of a Londoner—he may be from Manchester or Scotland or even some remote part of the world. This is the relationship of implication. Clearly these are all different logical relationships, and even if they are applied to the same situations the compound situations formed will be different.

For example, if A is all the prime numbers and B all numbers which when divided by 4 leave remainder 1, then 'A and B' means only those prime numbers which when divided by 4 leave remainder 1; 'A or B' means all prime numbers as well as any other numbers there might be which when divided by 4 leave remainder 1.

Bruner, Goodnow and Austin, in a recent research project at Harvard University, studied subjects' reactions to the different logical combinations of already established concepts. (8) In other words, the ingredients of the concepts were simple, so simple that everybody knew them. The investigation concerned the individual strategies by which the subject tried to discover the logical relationships, which were the only unknowns in the problems.

The procedure in most cases was the presentation of a number of cards. Each card had on it triangles, circles or squares, one or two or three of these; and each card was red or blue or green. So there were three variables—number, shape, and colour—each with three values. Then a concept such as red triangles was thought of by the experimenter, and the subject chose cards to which the experimenter answered either Yes or No: Yes if the card was red *and* had triangles on it, and No if not. Subjects were asked to find the concept in the least number of tries. Sometimes more variables were used, sometimes the number of choices was restricted. The problems were simple enough for mathematically ideal stragegies to be worked out, and the actual strageties observed were compared with these.

The reader will see how this procedure can easily be adapted to the examination of the formation of conjunctive, disjunctive and other concepts. For example, 'red triangles' is *conjunctive*, since the card has to be red and consist of triangles. On the other hand 'red *or* triangles' is a *disjunctive* concept, and any red card as well as any card with triangles on would have elicited a Yes from the experimenter.

The differences between learning-situations may not lie only in the logical structure of the learning tasks. Different individuals attack the same problem differently, and this was shown clearly in the Harvard research. The type of problem attacked may, in fact, influence the angle from which it is attacked, but it is even more certainly influenced by the type of thinking individuals habitually make use of. Sir Frederick Bartlett lists and examines a number of such different types of thinking, ranging from what he calls 'closed system thinking' to the utterly different thinking of the artist, which he aptly calls 'adventurous thinking'. (9)

The problem of learning is essentially how to find a kind of 'best fit' between the structure of the task and the structure of the person's thinking. For the process to be explained by any kind of intelligible theory, both these structures must be taken into account and at least some attempt made at quantitative description. This is, of course, a very difficult task and little is so far known, though a small beginning has been made in the author's recent monograph *Concept Formation and Personality*. (10)

The reader who wishes to go into the detail of the theory and of the experimental evidence available is referred to the monographs listed in the bibliography. There is no space in this short section to elaborate matters. Only the main conclusions and resulting hypotheses will be given, for on these the ensuing practical suggestions about mathematics-learning will be based.

We have already considered Piaget's three stages in the formation of a concept. To each of these corresponds a very different type of learning. To the preliminary or play stage corresponds a rather undirected activity, seemingly purposeless—the kind of activity that is performed and enjoyed for its own sake. It is this kind of behaviour that is usually described as play. In order to make play possible, freedom to experiment is necessary. This stage of concept-learning should therefore be as free as possible, with the ingredients of the concept available as play material. The second stage is more directed and purposeful but is characterized by lack of any clear realization of what is being sought. At this stage a certain degree of structured activity is desirable. How this is developed will depend on the structure of the concept as well as on the subject's particular way of thinking. Until more is known about these factors the safest procedure is the provision of a great number of experiences, of varying structure but all leading to the concept. The third stage must provide adequate practice for the fixing and application of the concepts that have been formed. The games that are played during these stages will be referred to as

(a) preliminary games
(b) structured games
(c) practice games.

This classification is, of course, relative to a given concept. Clearly a practice game for one concept can act as a preliminary game for a later concept. It is, however, important not to use practice games as preliminary games for the *same* concept—a common error in infant schools, where children are often expected to learn from games which they cannot really play without already knowing what they are supposed to be learning. It is also important to be aware when a child is passing from one stage to another, so that appropriate experiences can be provided to keep pace with the changing situation.

Turning to Bruner's research, we find that one of the results relevant to mathematics-learning is the difficulty of disjunctive concepts. We must remember that Bruner's work is based almost entirely on logical dissection of the situation, and a logical disjunction stripped bare of its mathematical context is much more difficult than the same thing imagined as part of a structure in the process of being built up. For example, to pass from the statement

$$(x - 2) \times (x - 1) = 0$$
to *either* $(x - 2) = 0$ *or* $(x - 1) = 0$

is a very difficult task out of the mathematical context.

It will be shown that there is a constructive way of arriving at this conclusion which makes it considerably easier and so accessible to much younger children. Mathematical learning being pre-eminently one of construction of predicates followed only afterwards by a critical, i.e. logical, examination of what has been constructed, we cannot expect the Bruner type of study to illuminate more than small parts of the structure. The building up of whole pieces of the structure is not a logical operation and cannot therefore be examined in Bruner's analytical way. Bartlett gets rather nearer when he implies that thinking tends at times to break out of the tight, local systems to which some would confine it. Formal, logical thinking is circumscribed, and the Bruner type of concept-formation is one that remains within such carefully circumscribed and determined limits. This is not to say that such studies are not useful; they are merely insufficient to describe the more constructive type of mathematical thinking. It might be relevant to ask what would happen if the limits within which this closed type of thinking proceeds were to be considerably extended. Suppose we went on increasing the number of variables and the number of values each variable was allowed to assume, what would happen to the strategies? What happens when the amount of analysis needed to assess a situation from a logical point of view gets beyond the subject's capacity to hold all the possibilities before him? He must either give up or do something else. If he does not give up, it is very probable that something like the Bartlett type of adventurous thinking takes place. The essence of this kind of thinking is that the subject keeps before him what Bartlett calls a 'standard' towards which he works. An artist lives up to his own standard or he ceases to be one. An artist could not possibly analyse his problem logically: the number of possibilities is too enormous. It is small wonder that he has recourse to quite different thought-processes. When an artist has painted a picture which conforms to his standard, perhaps he is once again not so far from the mathematician who has constructed a new predicate.

If this is so, then mathematical thinking will need the sort of investigation which catches the constructive process while it is still going on. The problem is to devise standard mathematical situations in which this adventurous kind of thinking can still take place. This was done by the present author in a series of experiments which have opened up a new method of investigating mathematical processes at work. Some tasks were designed to be loaded more heavily on the constructive side: that is to say, many predicates had to be formed on top of one another. These tasks were at the same time logically simple, requiring very little analysis. Other tasks were designed whose logical structure was complex but where not much construction was required. In this way it was hoped to throw some light on the functioning of different types of thinking by noting the effectiveness with which different subjects coped with these situations.

It was found, and has since been amply confirmed in work with school children attacking mathematical tasks, that subjects varied in the extent to which they were able or willing to engage in analytical (logical) and constructive thinking respectively. It was also quite clear that children developed constructive thinking long before analytical thinking. So, in devising mathematical learning-situations by means of apparatus, it has to be remembered that although children may not be ready to make logical judgments, they are well able to build mathematical concepts much earlier than has been thought possible. The logical exploration of what they have built will naturally follow, but perhaps years later.

Let us now go into some detail about the *content* of mathematical learning. We know that the three stages of growth are necessary before a mathematical predicate or concept becomes fully operational. How can we accelerate the growth of the concepts by putting the most suitable experiences in the children's way? A mathematical concept usually contains a certain number of variables and it is the constancy of the relationship between these, while the variables themselves vary, that constitutes the mathematical concept. (5) (12) To give the maximum amount of experience, structured so as to encourage the growth of the concept, it seems *a priori* desirable that *all* possible variables should be made to vary while keeping the concept intact. For example, with the concept of a parallelogram we can vary the shape by varying the angles and the lengths of the opposite sides; we can vary the position, as long as we keep the opposite sides parallel. Clearly a set of congruent parallelograms placed in the same position would not be a suitable set of experiences for the growth of the concept. We might formulate this by saying that as many variables as possible should be made to vary so as to provide optimum experience in concept growth.

We must next examine the problem of the choice of structure for the actual conceptual content of the task. The result of the experimental work so far carried out suggests that the logical complexity should be kept to a minimum. If there is a choice between a constructively loaded task and an analytically loaded one, the former will almost certainly be more suitable, especially when the children are young. The analytical way of critically sizing up a situation is a much more mature way of thinking and very seldom occurs in children before the age of 12. After this age, analytical tasks (such as proofs) begin to be appreciated. These can be gradually introduced, always provided that the mathematical construction is there so that there is something to analyse.

Having decided on the structuring of the task, how are we going to allow for all the possible individual differences that exist in coping with the formation of the *same* concept? As I have said before, in our present state of knowledge—or rather, lack of it—the only way to do this is to play as many variations as possible in different media on the same conceptual theme. This can be done by providing tasks which *look* quite different but have essentially the same conceptual structure. In other words, we can vary the perceptual representation, keeping the conceptual structure constant. For example, parallelograms can be drawn on paper, they can be made out of two congruent wooden triangles, they can be traced out with pegs on a pegboard, they can be found in wallpaper patterns and so on. . . . Children learn what there is in common among these different representations, and it is this common feature which *is* the mathematical concept.

We can sum up as follows:

1. *Dynamic Principle.* Preliminary, structured and practice games must be provided as necessary experiences from which mathematical concepts can eventually be built, so long as each type of game is introduced at the appropriate time.

Although while children are young these games must perforce be played with concrete material, mental games can gradually be introduced to give a taste of that most fascinating of all games, mathematical research.

2. *Constructivity Principle*. In the structuring of the games, construction should always precede analysis, which is almost altogether absent from children's learning until the age of 12.

3. *Mathematical Variability Principle*. Concepts involving variables should be learnt by experiences involving the largest possible number of variables.

4. *Perceptual Variability Principle*. To allow as much scope as possible for individual variations in concept-formation, as well as to induce children to gather the mathematical essence of an abstraction, the same conceptual structure should be presented in the form of as many perceptual equivalents as possible.

It will be clear that the kind of mathematics-learning described here is not one that very often takes place in the conventional class-lesson. This is not to say that it never takes place. There are some children who are able to gather a degree of abstraction from very scanty experience, something which is impossible for most; to this may be added the chance that the teacher's type of thinking may coincide with a pupil's type of thinking, in which case the 'explanations' are more likely to strike home. But on the whole the kind of mathematics that is learnt is not of the kind which is a series of superimposed concepts, each gleaned from personal experience, through the psychodynamical process so aptly described by Piaget. The kind of mathematics that is learnt is of an associative kind; that is to say, children associate certain situations with certain processes, and carry out the processes every time they find themselves in the situations with which the processes have been associated. If the situations are slightly varied, as in the case of a problem formulated slightly differently, or often even through the use of different letters, a completely novel situation is created for the child. As transfer does not take place (since there is no general apprehension of the situation), the processes are not carried out, or the 'wrong' ones are carried out, and so the answer will be 'wrong'. In order to establish a learning-situation which will fit the fundamental requirements of mathematical learning as defined here, quite a different classroom organization and communication system is necessary.

To allow for children's individual differences most of the learning should take place individually or in small groups of twos and threes. It is not likely that more than three children will work at the same pace and in roughly the same way. This means that all the information cannot come from the teacher, as he simply would not have time to go round and 'teach' perhaps forty different children separately, all possibly at different stages. There must be other sources of information in the classroom, as well as places to find out what to do next and, if possible, places to check the accuracy or correctness of an answer, if this is appropriate. Accordingly, a system of assignment cards should be available from which the children can work. These should be arranged both in series—building up a concept by a series of related tasks—and in parallel—presenting the same conceptual idea in different material. Preferably, the cards should offer a degree of choice, and certainly great variety, in accordance with the two variability principles. In order to make the learning as constructive as possible, a considerable amount of concrete material will be necessary. The manipulation of this material, according to instructions given on the cards, will lead the children through the appropriate experiences, taking them from concept to concept and helping them to build up the conceptual structure of mathematics in their minds. The structured games, leading to the formation of abstract concepts, should be followed by problem exercises, as practical and meaningful as possible (not n taps all running at different rates into an enormous bath), to make sure that the concepts are truly operational before another cycle of concept-formation is allowed to begin.

. . . It goes without saying that an authoritarian attitude would not be helpful in a learning-situation of this kind. The essence of a creative learning-situation is keenness to inquire, and authoritarianism does not foster a spirit of inquiry. The teacher in charge has quite a different role to play. First, he must see to it that the lines of communication from the sources of information to the children are kept open. Often the wording of the cards creates a stumbling-block, or a child may be trying a card for which he is not ready, in the sense that he has not mastered the constituent concepts necessary. A re-drafting of the card or a word or two to the class as a whole will be necessary in the first case; and the introduction of the child to further experiences to strengthen the concepts he is learning, in the second case. The dynamic equilibrium of this kind of learning can be very delicate, and a cross word or a disapproving tone of voice may spoil the learning of a child for the rest of the lesson. Teachers in charge of such classes act as counsellors and helpers in the children's own efforts to grapple with the problems in front of them.

A conventionally trained teacher might wonder how all this work is kept going without the force of the authoritative figure at the head of the classroom. Will the children want to do all this, if they are not somehow made to? This is the vexed problem of motivation. Can you have a creative learning-situation in which the discipline is largely maintained by the spirit of inquiry—this being generated by interest in the tasks in which the children are engaged? In other words, does self-motivating activity lead to self-discipline? The answer is not entirely simple. It is not true to say that the teacher in charge can abstract himself from a feeling of being responsible for the conduct of the class. The children must feel that the teacher in charge is really in charge, that he will help them and control them if need be. But this task for the teacher is made much easier by the interest of the tasks. If a teacher administering a creative learning-scheme is himself convinced of its rightness, his enthusiasm will be caught by the children and the problem of class discipline will dwindle. As in most things, the proof of the pudding is in the eating. We have sufficient evidence now to say that any good teacher who has an easy relationship with his pupils is perfectly able to handle the disciplinary problems of this kind of situation if he is able to handle the other, more usual type of situation. The joy with which a creative kind of algebra lesson is greeted, the voluntary performance of mathematical activity in free time or time of free choice, are sufficient indications of the self-motivating character of the creative learning-situation.

It is not proposed to go into any theories of motivation, for which such a volume as this is clearly not the place. On the other hand, the problem had to be mentioned, because the social situation created by this type of mathematical learning is so different from the conventional class-lesson. In short, teachers who intend to use the method would do well to consider whether they are in agreement with the kind of social situation which its implementation implies. An attitude of sympathy and love for the pupils is essential, and an attitude of humility before the unfolding of children's powers of thinking is extremely desirable, as well as a general non-dogmatic approach.

Such attitudes are not acquired in a short time, and I am perfectly well aware that this human element may become a fundamental limiting factor. The teacher who is finally wedded to playing an authoritarian role in a formal class-teaching situation is not going to be moved by anything contained in these pages. The teacher, on the other hand, whose first instinct is sympathy with the child, and not power and authority over him, may find the additional demands on his sensitivity a challenge. Such a teacher will not find it difficult to learn that if a child makes a mistake it is better to suggest another task, which will make the child aware of his mistake, than to put a cross against the result. He will not find it difficult to learn to consider the merits of ways different from his own produced by the children, or to realize that the standardizing of a process may not be the only way of teaching its efficient performance. In short, he will readily learn in a hundred other ways to draw the children into a mutually creative situation in which everyone plays a positive part, not least of all the teacher himself.

References

1. Piaget, J. (and others) *L'enseignements mathématiques* (Neuchatel & Paris, 1955).

2. Piaget, J. *The child's conception of number* (London, 1952).

3. Hadamard, J. *An essay on the psychology of invention in the mathematical field* (New York, 1954).

4. Waismann, F. *Introduction to mathematical thinking: the formation of concepts in modern mathematics* (London, 1951).

5. Wertheimer, M. *Productive thinking* (New York, 1945).

6. Rokeach, M. *A study in religious and political dogmatism* Psychological Monographs (New York, 1956).

7. Adorno, T. W. (and others) *The authoritarian personality* (New York, 1950).

8. Bruner, J. S. (and others) *A study of thinking* (New York, 1956).

9. Bartlett, Sir Frederick. *Thinking* (London, 1958).

10. Dienes, Z. P. *Concept formation and personality* (Leicester, 1959).

11. Dienes, Z. P. *On the growth of mathematical concepts in children through experience* (In *Educational Research*, Vol. II, No. 1, November, 1959).

12. Gattegno, C., Servais, W., Castelnuovo, E., Nocolet, J. L., Fletcher, T. J., Motard, L., Campedelli, L., Biguenet, A., Peskett, J. W., Puig Adam, P. *Le materiel pour l'enseignement des mathématiques* (Neuchatel & Paris, 1958).

13. Stern, C. *Children discover arithmetic* (London, 1953).

14. Gattegno, C. and Cuisenaire, G. *Numbers in colour* (London, 1954).

15. Dienes, Z. P. *Introduction to the use of the multibase arithmetic blocks* (M.A.B.) and *Introduction to the use of the algebraic experience material* (A.E.M.) (E.S.A. School Materials Division, Pinnacles, Harlow, Essex).

Exercises: A Theory of Mathematics Learning

1. What is Dienes' view on the role of stimulus-response learning theory in the learning of mathematics? Do you agree? Argue for your position.

2. Dienes writes: "If this kind of endless open thinking is the essence of mathematical thinking, then it is clearly radically different from the more mundane, everyday type of thinking. Psychologists who have studied the problems of learning and thinking have seldom been mathematicians; perhaps this is the reason why no adequate theory has so far been proposed to account for the kind of learning that takes place in this rather special field." Would Dienes attack the work of Gagné? Illustrate your position.

3. Dienes appears to be using the ideas of Piaget to support his discussion. Do you see the relationship that Dienes claims is there? Comment.

4. Dienes sums up his position by citing four principles: Dynamic Principle, Constructivity Principle, Mathematical Variability Principle, and Perceptual Variability Principle. Show by a specific example from the classroom how each of these principles should be applied (according to Dienes).

5. What is your opinion of the Dienes approach? How does it relate to the work of Bruner, Brownell, Davis, Gagné, and Piaget?

Facilitating Meaningful Verbal Learning in the Classroom

David P. Ausubel

In mathematics, as in other scholarly disciplines, pupils acquire subject-matter knowledge largely through meaningful reception learning of presented concepts, principles, and factual information. In this paper, therefore, I first propose to distinguish briefly between reception and discovery learning, on the one hand, and between meaningful and rote learning, on the other. This will lead to a more extended discussion of the nature of meaningful verbal learning (an advanced form of meaningful reception learning) and the reasons it is predominant in the acquisition of subject matter; of the manipulable variables that influence its efficiency; and of some of the hazards connected with its use in the classroom setting.

Reception Versus Discovery Learning

The distinction between reception and discovery learning is not difficult to understand. In reception learning the principal content of what is to be learned is presented to the learner in more or less final form. The learning does not involve any discovery on his part. He is required only to internalize the material or incorporate it into his cognitive structure so that it is available for reproduction or other use at some future date. The essential feature of discovery learning, on the other hand, is that the principal content of what is to be learned is not given but must be discovered by the learner before he can internalize it; the distinctive and prior learning task, in other words, is to discover something. After this phase is completed, the discovered content is internalized just as in reception learning.

Meaningful Versus Rote Learning

Now this distinction between reception and discovery learning is so self-evident that it would be entirely unnecessary to belabor the point if it were not for the widespread but unwarranted belief that reception learning is invariably rote, and that discovery learning is invariably meaningful. Actually, each distinction constitutes an entirely inde-

Reprinted from *The Arithmetic Teacher,* vol. 15 (February 1968) 126-32. © 1968 by the National Council of Teachers of Mathematics. Used by permission.

pendent dimension of learning. Thus reception and discovery learning can each be rote or meaningful, depending on the conditions under which learning occurs. In *both* instances meaningful learning takes place if the learning task is related in a nonarbitrary and nonverbatim fashion to the learner's existing structure of knowledge. This presupposes 1) that the learner manifests a *meaningful learning set*, that is, a disposition to relate the new learning task nonarbitrarily and substantively to what he already knows, and 2) that the *learning task is potentially meaningful* to him, namely, relatable to his structure of knowledge on a nonarbitrary and nonverbatim basis. The first criterion, nonarbitrariness, implies some plausible or reasonable basis for establishing the relationship between the new material and existing relevant ideas in cognitive structure. The second criterion, substantiveness or nonverbatimness, implies that the potential meaningfulness of the material is never dependent on the exclusive use of particular words and no others, i.e., that the same concept or proposition expressed in synonymous language would induce substantially the same meaning.

The significance of meaningful learning for acquiring and retaining large bodies of subject matter becomes strikingly evident when we consider that human beings, unlike computers, can incorporate only very limited amounts of arbitrary and verbatim material, and also that they can retain such material only over very short intervals of time unless it is greatly overlearned and frequently reproduced. Hence, the tremendous efficiency of meaningful learning as an information-processing and -storing mechanism can be largely attributed to the two properties that make learning-material potentially meaningful.

First, by nonarbitrarily relating potentially meaningful material to established ideas in his cognitive structure, the learner can effectively exploit his existing knowledge as an ideational and organizational matrix for the understanding, incorporation, and fixation of new knowledge. Nonarbitrary incorporation of a learning task into relevant portions of cognitive structure, so that new meanings are acquired, also implies that the newly learned meanings become an integral part of an established ideational system; and because this type of anchorage to cognitive structure is possible, learning and retention are no longer dependent on the frail human capacity for acquiring and retaining arbitrary associations. This anchoring process also protects the newly incorporated material from the interfering effects of previously learned and subsequently encountered similar materials that are so damaging in rote learning. The temporal span of retention is therefore greatly extended.

Second, the substantive or nonverbatim nature of thus relating new material to and incorporating it within cognitive structure circumvents the drastic limitations imposed by the short item and time spans of verbatim learning on the processing and storing of information. Much more can obviously be apprehended and retained if the learner is required to assimilate only the substance of ideas rather than the verbatim language used in expressing them.

It is only when we realize that meaningful learning presupposes only the two aforementioned conditions, and that the rote-meaningful and reception-discovery dimensions of learning are entirely separate, that we can appreciate the important role of meaningful reception learning in classroom learning. Although, for various reasons, rote reception learning of subject matter is all too common at all academic levels, this need not be the case if expository teaching is properly conducted. We are gradually beginning to realize not only that good expository teaching can lead to meaningful reception learning but also that discovery learning or problem solving is no panacea that guarantees meaningful learning. Problem solving in the classroom can be just as rote a process as the outright memorization of a mathematical formula without understanding the meaning of its component terms or their relationships to each other. This is obviously the case, for example, when students simply memorize rotely the sequence of steps involved in solving each of the "type problems" in a course such as algebra (without having the faintest idea of what they are doing and why) and then apply these steps mechanically to the solution of a given problem, after using various rotely memorized cues to identify it as an exemplar of the problem type in question. They get the right answers and undoubtedly engage in discovery learning. But is this learning any more meaningful than the rote memorization of a geometrical theorem as an arbitrary series of connected words?

In meaningful classroom learning, the balance between reception and discovery learning tends, for several reasons, to be weighted on the reception side: First, because of its inordinate time-cost, discovery learning is generally unfeasible as a *primary* means of acquiring large bodies of subject-matter knowledge. The very fact that the accumulated discoveries of millennia can be transmitted to each new generation in the course of childhood and youth is possible only because it is so much less time-consuming for teachers to communicate and explain an idea meaningfully to pupils than to have them rediscover it by themselves. Second, although the extent and complexity of meaningful reception learning in pure verbal form is seriously limited in pupils who are either cognitively immature in general or unsophisticated in a particular discipline, the actual process of discovery per se is never required for the meaningful acquisition of knowledge. Typically it is more efficient pedagogy to compensate for such deficiencies by simply incorporating concrete-empirical props into expository teaching techniques. Finally, although the development of problem-solving ability as an end in itself is a legitimate objective of education, it is less central an objective than that of learning the subject matter. The ability to solve problems calls for traits such as flexibility, originality, resourcefulness, and problem-sensitivity that are not only less generously distributed in the population of learners than is the ability to understand and retain verbally presented ideas but are also less teachable. Thus relatively few good problem solvers can be trained in comparison with the number of persons who can acquire a meaningful grasp of various subject-matter fields.

The Nature of Meaningful Reception Learning

Like all learning, reception learning is meaningful when the learning task is related in nonarbitrary and nonverbatim fashion to relevant aspects of what the learner already knows. It follows, therefore, from what was stated above that the first precondition for meaningful reception learning is that it take place under the auspices of a meaningful learning set. Thus irrespective of how much potential meaning may inhere in a given proposition, if the learner's intention is to internalize it as an arbitrary and verbatim series of words, both the learning process and the learning outcome must be rote or meaningless.

One reason why pupils commonly develop a rote-learning set in relation to potentially meaningful subject matter is that they learn from sad experience that substantively correct answers lacking in verbatim correspondence to what they have been taught receive no credit whatsoever from certain teachers. Another reason is that because of a generally high level of anxiety or because of chronic failure experience in a given subject (reflective, in turn, of low aptitude or poor teaching), they lack confidence in their ability to learn meaningfully, and hence they perceive no alternative to panic apart from rote learning. This phenomenon is very familiar to mathematics teachers because of the widespread prevalence of "number shock" or "number anxiety." Lastly, pupils may develop a rote-learning set if they are under excessive pressure to exhibit glibness, or to conceal rather than admit and gradually remedy original lack of genuine understanding. Under these circumstances it seems both easier and more important to create a spurious impression of facile comprehension by rotely memorizing a few key terms or sentences than to try to understand what they mean. Teachers frequently forget that pupils become very adept at using abstract terms with apparent appropriateness — when they have to — even though their understanding of the underlying concepts is virtually nonexistent.

The second precondition for meaningful reception learning—that the learning task be potentially meaningful or nonarbitrarily and substantively relatable to the learner's structure of knowledge—is a somewhat more complex matter than meaningful learning set. At the very least it depends on the two factors involved in establishing this kind of relationship, that is, on the nature of the material to be learned and on the availability and other properties of relevant content in the particular learner's cognitive structure. Turning first to the nature of the material, it must obviously be sufficiently plausible and reasonable that it could be related on a nonarbitrary and substantive basis to *any* hypothetical cognitive structure exhibiting the necessary ideational background. This is seldom a problem in school learning, since most subject-matter content unquestionably meets these specifications. But inasmuch as meaningful learning or the acquisition of meanings takes place in *particular* human beings, it is not sufficient that the learning task be relatable to relevant ideas simply in the abstract sense of the term. It is also necessary that the cognitive structure of the *particular* learner include relevant ideational content to which the learning task can be related. Thus, insofar as meaningful learning outcomes in the classroom are concerned, various properties of the learner's cognitive structure constitute the most crucial and variable determinants of potential meaningfulness. These properties will be considered briefly but systematically in the following section.

At this point it is important to appreciate that the idiosyncratic nature of each learner's cognitive structure implies that the meanings he acquires from any potentially meaningful learning task must necessarily be idiosyncratic in nature. In fact, it could hardly be an overstatement of the case to say that the extent to which learning is meaningful largely depends on how idiosyncratic it is — that is, on how intimately the objective content of the learning task can be incorporated into the distinctively idiosyncratic aspects of the learner's relevant cognitive structure. Thus in a very real sense the meaningfulness of reception learning is in large measure a function of how actively and energetically a given pupil endeavors to translate new propositions into terminology consistent with his particular vocabulary and ideational background, and how self-critical he is in judging whether this goal has been accomplished. The main danger relative to meaningful reception learning is not so much that the learner will frankly adopt a rote-learning set but that he will be insufficiently energetic in reformulating presented propositions so that they have real meaning for him in terms of his own structure of knowledge, and that he will then delude himself and his teachers into believing that the resulting empty, vague, or imprecise verbalisms are genuinely meaningful.

The Role of Cognitive Structure Variables in Meaningful Verbal Learning

Since, as suggested above, the potential meaningfulness of a learning task depends on its relatability to a particular learner's structure of knowledge in a given subject-matter area or subarea, it follows that *cognitive structure itself*, that is, both its substantive content and its major organizational properties, should be the principal factor influencing meaningful reception learning and retention in a classroom setting. According to this reasoning, it is largely by strengthening salient aspects of cognitive structure in the course of prior learning that new subject-matter learning can be facilitated. In principle, such deliberate manipulation of crucial cognitive structure variables — by shaping the content and arrangement of antecedent learning experience — should not meet with undue difficulty. It could be accomplished 1) *substantively*, by using for organizational and integrative purposes those unifying concepts and principles in a given discipline that have the greatest inclusiveness, generalizability, and explanatory power, and 2) *programmatically*, by employing optimally effective methods of ordering the sequence of subject matter, constructing its internal logic and organization, and arranging practice trials.

Both for research and for practical pedagogic purposes it is important to identify those manipulable properties or variables of existing cognitive structure that influence the meaningful reception learning of subject-matter knowledge. On logical grounds, three such variables seem self-evidently significant: 1) the *availability* in the learner's cognitive structure of relevant and otherwise appropriate ideas to which the new learning material can be nonarbitrarily and substantively related, so as to provide the kind of anchorage necessary for the incorporation and long-term retention of subject matter; 2) the extent to which such relevant ideas are *discriminable* from similar new ideas to be learned so that the latter can be incorporated and retained as separately identifiable entities in their own right; and 3) the *stability* and *clarity* of relevant anchoring ideas in cognitive structure, which affect both the strength of the anchorage they provide for new learning material and their degree of discriminability from similar new ideas in the learning task.

Availability of Relevant Anchoring Ideas in Cognitive Structure

One of the principal reasons for rote or inadequately meaningful learning of subject matter is that pupils are frequently required to learn the specifics of an unfamiliar discipline before they have acquired an adequate foundation of relevant and otherwise appropriate anchoring ideas. Because of the unavailability of such ideas in cognitive structure to which the specifics can be nonarbitrarily and substantively related, the latter material tends to lack potential meaningfulness. But this difficulty can largely be avoided if the more general and inclusive ideas of the discipline, that is, those which typically have the most explanatory potential, are presented first and are then progressively differentiated in terms of detail and specificity. In other words, meaningful reception learning and retention occur most readily and efficiently if, by virtue of prior learning, general and inclusive ideas are already available in cognitive structure to play a *subsuming* role relative to the more differentiated learning material that follows. This is the case because such subsuming ideas when established in the learner's structure of knowledge 1) have maximally specific and direct relevance for subsequent learning tasks, 2) possess enough explanatory power to render otherwise arbitrary factual detail potentially meaningful (i.e., relatable to cognitive structure on a nonarbitrary basis), 3) possess sufficient inherent stability to provide the firmest type of anchorage for detailed learning material, and 4) organize related new facts around a common theme, thereby integrating the component elements of new knowledge both with each other and with existing knowledge.

One of the more effective strategies that can be used for implementing the principle of progressive differentiation in the arrangement of subject-matter content involves the use of special introductory materials called "organizers." A given organizer is introduced in advance of the new learning task per se; is formulated in terms that, among other things, relate it to and take account of generally relevant background ideas already established in cognitive structure; and is presented at an appropriate level of abstraction, generality, and inclusiveness to provide specifically relevant ideational scaffolding for the more differentiated and detailed material that is subsequently presented. An additional advantage of the organizer, besides guaranteeing the availability of specifically relevant anchoring ideas in cognitive structure, is that it makes explicit both its own relevance and that of the aforementioned background ideas for the new learning material. This is important because the mere availability of relevant anchoring ideas in cognitive structure does not assure the potential meaningfulness of a learning task unless this relevance is appreciated by the learner. Lastly, it is desirable not only for the material within each topic to become progressively more differentiated — both by using organizers and by proceeding from subtopics of greater to lesser inclusiveness in the learning material itself — but also to follow the same organizational plan in ordering the sequence of the various topics comprising a given course of study.

It is also possible in subject-matter learning to capitalize on the availability in cognitive structure of relevant anchoring ideas reflective of prior incidental experience or nonverbal learning. This is the underlying rationale for the widely accepted pedagogic practice of proceeding from intuitively familiar to intuitively unfamiliar topics in sequencing subject matter, thereby using previously acquired intuitive principles or general background as a foundation for learning less familiar material.

Finally, the availability of relevant anchoring ideas for use in meaningful verbal learning may be maximized by taking advantage of natural sequential dependencies among the component divisions of a particular discipline, i.e., of the fact that the understanding of a given topic often logically presupposes the prior understanding of some related topic. Thus, by arranging the order of topics and subtopics in a given subject-matter field as far as possible in accordance with these sequential dependencies, the learning of each unit, in turn, not only becomes an achievement in its own right but also constitutes specifically relevant ideational scaffolding for the next item in the sequence.

Consolidation of Anchoring Ideas

The sequential organization of subject matter naturally assumes that any given step in a particular sequence is always clear and stable before the next step is presented. If this is not the case, the anchorage it furnishes for all subsequent steps is insecure, and their learning and retention are accordingly jeopardized. Hence, new material in the sequence should never be introduced until the preceding step is thoroughly mastered. Such mastery, of course, can be achieved only through adequate and differential practice, review, testing, and feedback which provide the necessary confirmation, clarification, and correction required for the effective consolidation of meaningfully learned material. Consolidation also facilitates meaningful verbal learning by increasing the discriminability of previously learned material from similar new learning tasks—those that are sequentially dependent and those that are not.

Discriminability of Learning Material from Established Ideas

This brings us to the role of discriminability in meaningful verbal learning. It is self-evident that before new ideas can be meaningfully learned, they must be adequately discriminable from similar established ideas in cognitive structure. If the learner cannot discriminate clearly, for example, between new idea A' and previously learned idea A, then A' enjoys relatively little status as a separately identifiable meaning in its own right, even at the very onset of its incorporation into cognitive structure. In addition, if new meanings cannot be readily distinguished from previously learned established meanings, they can certainly be adequately represented by them for memorial purposes, and thus they tend to be reduced to the latter even more rapidly than is typically the case in the retention of new meanings. In other words, only discriminable variants of established ideas in cognitive structure have long-term retention potentialities.

Thus, in learning situations where new ideas are introduced that are similar to previously learned ideas and hence confusable with them, it is advisable, by means of a procedure known as *integrative reconciliation*, to point out *explicitly* the basic similarities and differences between them. This practice integrates knowledge by specifically identifying the commonalities underlying similar ideas; by preventing artificial compartmentalization and the proliferation of separate terms for concepts that are basically identical except for contextual usage; and, most important, by sharply delineating in what ways similar but not identical ideas are actually different. Failure to specify such relationships between previously acquired and later-appearing subject-matter content, that is, treating the latter content in self-contained fashion without explicitly attempting to reconcile it with the former, assumes rather unrealistically that students will adequately perform the necessary cross-referencing.

Where necessary, organizers can also further the goal of integrative reconciliation by explicitly delineating the essential similarities and differences between the new subsuming concepts and principles to be learned and similar established ideas in cognitive structure. By so enhancing the discriminability of the newly introduced anchoring ideas, such organizers enable the learner to grasp the more differentiated aspects of the new learning task with many fewer ambiguities and misconceptions than would otherwise be possible. This differentiated material is also retained longer both because it is learned more clearly in the first place (by virtue of the greater discriminability of the new anchoring ideas under which it is subsumed) and because more discriminable subsumers are themselves more stable and hence better able to provide secure anchorage.

Exercises: Facilitating Meaningful Verbal Learning in the Classroom

1. Is there a relationship between Ausubel's emphasis on advance organizers and Gagné's hierarchies?

2. Ausubel lists several reasons why students commonly develop a rote-learning set. Make a list of ways teachers can encourage the development of meaningful learning sets.

3. Does Ausubel's emphasis on advance organizers reflect a particular philosophy about the nature of knowledge itself? The concept of advance organizer seems to be most useful in cases where one is learning a large body of facts. Its use in learning relationships, such as those of an algebraic group is much less clear. Why might this be so?

4. Many primary teachers introduce multiplication in terms of repeated addition. How does this approach act as an advance organizer? Does it fail to meet any of Ausubel's criteria for advance organizers.

5. Are advance organizers related to a student desire for "relevance"? If so, how? If not, how do they differ?

6. The preceding article of Gagné's discusses learning as information processing. How might the idea of an advance organizer be considered within the information processing model of human learning?

7. Provide an example of an advance organizer for long division, slope, inverse operations, and the Pythagorean theorem.

8. Can Dienes' constructivity principle be used as a mechanism to assure that verbal learning is meaningful? Explain.

Learning and Proficiency in Mathematics

Robert M. Gagné

In this report I should like to tell you about some of the ideas and products which have resulted from a collaboration of experimental psychologists, mathematicians, and mathematics teachers. The project which I have directed at Princeton University has been involved in such a collaboration with the University of Maryland Mathematics Project, which, as you may know, has as its Director, Dr. John Mayor, and as Associate Director, Mrs. Helen Garstens.

What we have done together is to develop a method of conducting experimental studies of the learning of mathematics. We have used this method to investigate the action of several factors in the learning situation and to verify their effects on mathematics learning.

What is the method we have developed and used? How has it worked out in revealing the factors at work in learning? And what kinds of results does it lead to, with their implications? Perhaps it should be said, at the outset, that my description of this method is intended to stimulate thinking and discussion about the process of instruction as an investigable set of events. The major purpose here is to present a viewpoint about research and its relation to the process of instruction.

The Method

The primary method employed involved the use of what are called "programmed learning" materials. As readers of the literature on teaching machines know, such materials are designed to present information to the learner, and he is required to make a response to it by filling in a blank or answering a question [1].* Once he has done this, an answer frame is exposed which informs him of the correct response; he then proceeds to the next frame, and so on throughout the program.

In the studies being described, learning programs were devised by reproducing typed frames of information on index cards, or on half-size sheets of paper, and assembling them into booklets of convenient size having loose hinges which permitted the pages to be flipped over easily. The answers to the questions posed by the frames were printed on the back of the cards. Generally, students were instructed to turn back the card and read the frame again whenever they had made an incorrect

---

Reprinted from *The Mathematics Teacher*, vol. 56 (December, 1963): 620-626. © 1963 by the National Council of Teachers of Mathematics. Used by permission.
*Numerals in brackets refer to the References at the end of this article.

response. Answers were recorded by the students on specially prepared answer sheets numbered to correspond with the frame member.

Using such materials, studies were conducted both in the laboratory with individual students one at a time, and in classrooms, where groups of students responded to the learning programs each in his individual manner. The setup used with individual learners consisted of a visible card file mounted on a stand, in which cards containing the individual frames were inserted. The materials used with students in classrooms were looseleaf booklets and an answer sheet. The frames of the learning program proceeded in a step-by-step fashion, each requiring a response on the part of the student.

When school groups were used, provision had to be made for the fact that different students would finish a booklet at different times, not only because of individual differences in learning rate, but also because the learning program itself was presented in experimentally different versions. This situation was handled by having each student, upon completion of a booklet, turn his attention to other work (unrelated to the program) which had previously been assigned by the regular classroom teacher. In this manner it was possible to have all students finish all booklets.

Content of learning programs

The learning programs employed in these studies had various mathematical contents keyed to the level of mathematical sophistication of the students. Generally speaking, topics for the programs were designed to be unfamiliar to the students, but involved the assumption of particular previous knowledge in each case. In some instances, programmed learning materials were based upon a topic occurring at a particular point in a mathematics textbook and were introduced at the normal time in the instructional sequence. In other cases, the materials were concerned with topics which the students had not yet reached in their normal curriculum. Specifically, studies used materials on the following topics: (1) deriving formulas for the sum of terms in number series [2], (2) "solving" simple algebraic equations [3], and (3) deriving definitions for, and performing additions of, integers [4]. In studies currently underway, materials on tangents of angles and on basic nonmetric geometry are being used.

Measuring student performance

Following the completion of learning programs, a number of measures of performance were employed to reveal what the students had learned. Typically, one of these was a test of performance designed to measure proficiency in the class of tasks specifically covered in the learning program. In most instances, a test of transfer was also employed, in the attempt to determine the extent of generalizability of what had been learned. Such tests presented problems which were new to the student and which belonged to a class of tasks other than that included in the learning program. For example, a test of transfer given following a learning program on addition of integers included problems on addition of rational numbers. Finally, a third type of measure was a test of subordinate knowledges, which was intended to reveal whether a student could or could not perform correctly each of the types of tasks contained within the learning program, but subordinate to the final task for which the program was intended.

Variables in learning programs

The basic situation with which we were dealing, then, was one in which a learner interacted with the material on a printed page. It is of some interest to consider what kinds of variables may be at work when an instructional process is generated in this way. What functions are being performed by these printed statements? What do they do for the learner?

As described more completely elsewhere [3], the functions which these statements seemed to us to be performing included the following: (1) they may define for the learner the general form of performance expected at the end of each subtopic; (2) they define unfamiliar words and symbols; (3) they may require the learner to recall certain subordinate knowledges he has previously learned; and (4) they "guide" his thinking about the new task, while encouraging discovery. It may be noted that in an experiment, any or all of these functions of frames may be manipulated in order to test their effects on the learning outcomes. For example, one can give more or less guidance to thinking within a learning program. Or one can vary the number of times a newly acquired task is recalled and thus make repetition (in this specific sense) an experimental variable.

But there is another way in which one can vary the frames of a learning program, and this may be the most important of all. This is by ordering the topics throughout the program. By "topic" is meant a distinguishable principle of knowledge which can govern the performance of a class of human tasks. When we consider a final performance to be learned (such as "adding integers"), we find that it can be analyzed into a number of subordinate topics which must first be mastered before the final task can be attained. These topics in turn depend upon the mastery of other subordinate topics. This kind of analysis, then, results in the identification of a hierarchy of subordinate knowledges, such as that shown in Figure 17. Each element of this subordinate knowledge is hypothesized to support the learning of each topic in the hierarchy to which it is connected by an arrow. That is, mastery of the subordinate knowledge is considered to be essential to the attainment of a related higher-level topic; learning of the latter cannot occur without it. In Figure 17, for example, a task like IVa, "using 0 as the additive identity" must be mastered (according to hypothesis) in order for IIIa, "stating and using the definition of addition of an integer and its additive inverse," to be learned. Similarly, IIIa must first be mastered before IIa can be achieved, and so on.

The implications of this hypothesis about topical order are easy to grasp. If an individual learner has achieved the subordinate knowledge represented by IVa (Fig. 1), his learning of IIa will be highly probable if he has also learned IIIa, and highly improbable if he has not. When translated into a formula for the design and construction of a learning program, or other instructional sequences, this means that the topic IVa must precede the topic IIIa, which in turn must precede the topic IIa, and so on. If the order is violated, or an intervening topic omitted, the acquisition of knowledge of any superordinate topic will be unsuccessful.

The method of making an analysis to arrive at the set of subordinate knowledges arranged in a hierarchy like that of this figure should be briefly mentioned. One begins with the terminal class (or classes) of tasks for which learning is being undertaken. For each of these, one asks the question, "What must the learner already know how to do, in order to achieve this (new) performance, assuming that he is to be given only instructions?" The latter part of this question assumes that the instructions will have the functions of frames previously mentioned, with the exception of task repetition. The answer to this question defines one or more elements of subordinate knowledge (at level I in Fig. 17). The question is then applied to each of these in turn, thus identifying the entire hierarchy. The process ends when one arrives at subordinate knowledge (like Va and Vb) which can be assumed to be possessed by every learner for whom the learning program is intended.

This description should not be taken to imply, however, that there is only one unique hierarchy appropriate to the learning of any particular final performance. This particular one is based upon a logical sequence developed in UMMaP materials; had we used an SMSG chapter, the hierarchy would perhaps have been somewhat different. And, of course, it would be perfectly possible to construct a very simple hierarchy for Task 2 (adding integers) based upon three computational "rules." The importance of such hierarchies is not that they can be uniquely determined, but rather that they depict a learning structure which, once defined, indicates steps in instruction which must be accomplished in a proper sequence in order to achieve the desired performance.

The Method in Use

The approach described separates the factors which potentially influence mathematics learning into two broad categories. The first of these may be called *instructional* variables, conceived as a set of functions that are performed in presenting *each* new topic, or new principle, to be learned. As previously mentioned, they include the functions of definition of new stimuli and terminal performances, recalling, guiding thinking, and task repetition. The second broad category is *topical order*, which pertains to the sequence of subordinate knowledge (topics) that must be acquired in order to achieve some final performance. Any instructional sequence, one form of which is a learning program, utilizes certain selected values and arrangements of variables from each of these classes.

How do we actually go about putting this method into use in performing experimental studies?

First, we have to analyze a subject to be taught, to reveal its knowledge structure. By so doing, we are defining the subordinate knowledge that students must possess in order to master the task.

Second, we construct a basic learning program having the characteristics I suggested. It has a *topic order* determined by the subordinate-knowledge hierarchy. In proceeding from one subtopic to the next, care

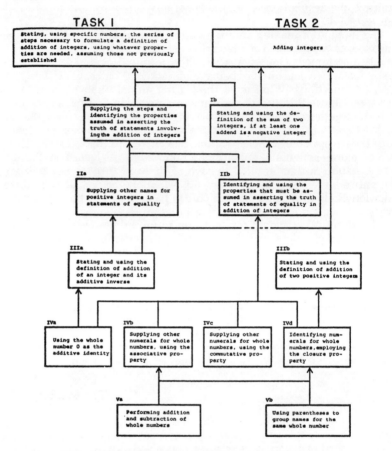

Figure 17. A Hierarchy of a Learning Program.

is taken that individual frames have performed the functions previously mentioned, namely, the definition of terminal performances and new terms, the activation of recall, and the guidance of thinking.

Having done this, we next introduce the *variations* in this basic program which we are interested in from an experimental standpoint. For example, if we want to study the effects of more or less guidance to thinking, we construct programs containing more or fewer frames performing this function. Or, if we want to study the effects of task repetition (because of its presumed influence on recallability), we construct several forms of programs providing variation in this variable.

Next, we administer these programs to classes of school children. Usually, they are divided into booklets in such a way that everyone finishes on each day.

Following this, we measure performance on a test designed to measure exactly what the program as a whole was designed to teach, no more and no less. In addition, if we are interested in the question of transfer of training, we administer a test involving some related but entirely new subject matter.

As a final step, we measure each person's subordinate knowledge — each one of the topics that is represented in the figure previously shown. If the student is not able to do the final task to perfection, we want to know why — which element of the subordinate knowledge he didn't have.

A word needs to be said about the scoring of the items used to test whether the individual could or could not perform the task set for him. These items were scored independently by two mathematics teachers, and their degree of agreement was virtually perfect. Of course, it should be remembered that this is somewhat unconventional "testing," which makes no attempt to measure degrees of knowledge, whatever that may be. The performance of each task was either wholly right or was considered wholly wrong.

Some Results

Let me describe briefly some of our results.

First, what about this knowledge hierarchy, and the topical order it determines? It will be recalled from the example shown in Figure 10, that a topic (i.e., an item of knowledge) at a particular level in the hierarchy may be supported by one or more topics at the next lower level. What is being predicted, then, is this: An individual will not be able to learn a particular topic if he has failed to achieve *any* of the subordinate topics that support it. This hypothesis can of course be tested at a number of different points throughout the hierarchy. With reference to the structure of Figure 10, our results showed that *for no topic were there more than three percent of instances contrary to this hypothesis* [4]. (We have no way of accounting for the three percent except as measurement error; however, we consider it to be gratifyingly small.)

These results imply something about individual differences. What they imply is that the most important difference among learners in their ability to perform a final task resides in their possession, or lack of possession, of this subordinate knowledge. Patterns of subordinate knowledge exhibited by successful and unsuccessful learners differ in quite predictable ways [3]. The unsuccessful learner, in progressing through the self-instructional program, effectively stops learning somewhere along the road through the knowledge hierarchy, and is unable to master any subordinate knowledge beyond that point.

As for a general aptitude like "intelligence," there is not surprisingly a moderately low degree of correlation with *rate* of progression through a learning program, and this correlation remains relatively constant. This correlation with rate, however, is to be distinguished from one with proficiency. The important thing suggested by our results is this: If all learners are allowed time to complete the program (as was true in our administration), and if this program is otherwise effective, their performance at the end comes to be *independent* of ability scores measured before the learning began, and highly dependent upon the specific subordinate knowledge they have learned. For example, in the study of learning about addition of integers, which turned out to have a moderately successful program (as indicated by average scores of 8.9 in achievement of the component knowledge, out of a total of 12), no significant relationships were found between mathematics grades before the experiment began and performance on the final task [4]. But there is a high correlation (.88) between the latter performance and the scores indicating the number of elements of subordinate knowledge learned.

What about instructional variables? As I have pointed out, it is possible to perform studies in which various modifications of these variables are incorporated into different forms of learning programs. One of our studies attempted to learn about the effects of two of these variables which can be manipulated by changing some features of the instructions provided in the frames of learning programs. One of these was task repetition — the number of additional examples given to the learner after he had first achieved each task of the knowledge hierarchy. A form of learning program characterized by "low repetition" gave one or two additional (varied) examples, while one having "high" repetition provided four or five times this number of examples, for each topic. It should be emphasized that what were given repetition were these subordinate tasks (see Fig. 1 for their names) in their terminal form, and not the remaining frames of the program.

Another factor subjected to variation was amount of guidance, represented by the number of frames taken to "instruct" the learner how to go from one topic to the next. One form of program used only a few frames to suggest a line of thinking to the learner ("low" guidance), while another used two or three times as many ("high" guidance). Neither of these forms could be said to "state the answer"; both required discovery on the part of the learner.

The results obtained on the effects of these variables can be rather simply stated. Neither repetition alone, nor guidance alone, had significant effects on the learning as measured by tests of final performance. When "high" repetition and "high" guidance were combined, however, this learning program produced a significantly higher number of successful learners than did the opposite combination of "low" repetition and "low" guidance [4], while other comparisons were not significantly different.

The results of course do not indicate a particularly strong effect by either of these learning variables. In fact, when the results of several of our studies are considered together, one does not gain the impression that instructional variables (such as guidance and repetition) have very pronounced effects upon the learning that takes place within the framework of an instructional program. This sends us back to a consideration of the contrastingly prominent effects of what may be called "content" variables, pertaining to the structure and organization of the knowledge being taught.

The most prominent implication of the results of these studies to date is that *acquisition of new knowledge depends upon the recall of old knowledge*. When stated in this way, the proposition seems to have few startling characteristics. In more specific form, however, it means that the learning of any particular capability requires the retention of other particular items of subordinate knowledge — not just any knowledge, or knowledge in some general sense. The learner acquires a new item of knowledge largely because he is able to integrate previously acquired principles into new principles, and he cannot do this unless he really knows these previously learned principles. The design of an instructional situation is basically a matter of designing a *sequence of topics.*

Our results imply that there are many, many specific sets of "readiness to learn." If these are present, learning is at least highly probable. If they are absent, learning is impossible. So, if we wish to find out how learning takes place, we must address ourselves to these specific readinesses to learn." If these are present, learning is at least highly probable. need to know a lot more about how they get established, and why they sometimes do not. The arrangement of the external conditions for learning is a matter of careful organization of the entities of knowledge, and their presentation in such a manner that no learner can help acquiring the new capabilities for achievement that we want to give him.

References

1. Lumsdaine, A. A., and Glaser, R. *Teaching Machines and Programmed Learning.* Washington, D.C.: National Education Association, 1960.

2. Gagné, R. M. "The Acquisition of Knowledge," *Psychol. Rev.,* LXIX (1962), 355-365.

3. _____ and Paradise, N. E. "Abilities and Learning Sets in Knowledge Acquisition," *Psychol. Monogr.,* No. 518 (1961), 75.

4. _____, Mayor, J. R., Garstens, H. L., and Paradise, N. E. "Factors in Acquiring Knowledge of a Mathematical Task," *Psychol. Monogr.,* No. 526 (1962), 76.

Exercises: Learning and Proficiency in Mathematics

1. Some would say that a Gagné learning hierarchy is appropriate for programmed learning but that it has little to do with ordinary classroom instruction. Do you agree? Why? Explain. Use examples from the classroom.

2. Examine the hierarchy given in Figure 17 carefully. Write a sample test item for each box in the hierarchy. Describe the possible problems that occur to you as you write the sample items.

3. Explain why there can be more than one hierarchy for a given task. Why isn't the learning hierarchy for a given task unique? Give two learning hierarchies for the same task.

4. What does Gagné say about individual differences? What individual differences among students do you think account for the major differences in learning in the mathematics classroom? Would the school counselors agree? Why?

5. Is Gagné saying anything more than that there is a logical order to mathematics? Justify your reaction with examples.

6. What implications do you see for your own teaching? If you see none, explain why. If you see applications, give examples.

7. List and discuss the factors you would consider in determining whether a student or class (5th grade level) were ready to learn a general algorithm for the addition of rational numbers.

8. The Gagné article in Unit 1 describes a hierarchy of learning types. This article details a hierarchy or flow chart of educational objectives for operations with integers. Are the two hierarchies comparable? Can one be translated directly into the other? Why?

Unit 4

Instructional Applications

Psychology focuses on the nature of learning. Instruction must go beyond this focus to encompass judgment of what should be learned. The articles for this section were selected to help classify the goals of instruction and to apply psychological judgment to the design of teaching strategies.

The section begins with two excerpts from taxonomies of objectives. The Bloom and Krathwohl classification schemes for outcomes of learning provide a contrast for the hierarchy of learning types developed by Gagné. These articles suggest the kind of balance teachers should seek in planning instructional goals. The Brownell article is an example of how psychological judgment is tempered by a feeling for the range and kind of objectives to be attained.

Interest in particular facets of how children learn has created concentrated emphases in mathematics education in the past. The next three articles are from an era of the sixties in which concern with discovery learning may have been predominant in curriculum design. Some educators were saying that if a concept was not readily discoverable, then it was not worth teaching. Davis presents a cogent analysis of discovery teaching and concludes with an analysis of goals of instruction in terms of discovery learning. A portion of the research base for discovery is developed by Hendrix in a manner yielding insight about learning in general. Ausubel's analysis of discovery learning is typical of what comes later in the era of a developing instructional approach. It is a reasoned, balanced presentation which attempts to marshall a variety of pertinent evidence while arguing for a reasonable balance in teaching.

This section closes with one man's attempt to balance knowledge of the psychology of learning with desired outcomes of learning in mathematics. Davis demonstrates an enviable eclecticism as he strives to develop a practical theory of instruction.

Condensed Version of the Taxonomy of Educational Objectives: Cognitive Domain

Introduction to Taxonomy of Educational Objectives: Cognitive Domain

What follows is a condensed version of a classic in education commonly referred to as "Bloom's Taxonomy." It is an attempt to classify cognitive educational objectives in a hierarchy from "low level" to "high level" objectives. Its greatest strength is that it helps one to focus on one's own objectives in the classroom and use the taxonomy as a checklist. Has something vitally important been forgotten? What objectives are most important to students? What are you doing in the classroom to meet these objectives?

Read the taxonomy with your own teaching in mind. Make up examples of each category from your own classroom. Challenge the taxonomy. Does it leave out something you think is important? Does it dwell on things you feel are unimportant. To evaluate a taxonomy you must use it. Does it work?

1.00 Knowledge

Knowledge, as defined here, involves the recall of specifics and universals, the recall of methods and processes, or the recall of a pattern, structure, or setting. For measurement purposes, the recall situation involves little more than bringing to mind the appropriate material. Although some alteration of the material may be required, this is a relatively minor part of the task. The knowledge objectives emphasize most the psychological processes of remembering. The process of relating is also involved in that a knowledge test situation requires the organization and reorganization of a problem such that it will furnish the appropriate signals and cues for the information and knowledge the individual possesses. To use an analogy, if one thinks of the mind as a file, the problem in a knowledge test situation is that of finding in the problem or task the appropriate signals, cues, and clues which will most effectively bring out whatever knowledge is filed or stored.

Benjamin S. Bloom, et al., "Condensed Version of the Taxonomy of Educational Objectives," *Taxonomy of Educational Objectives, Handbook I: Cognitive Domain.* (New York: David McKay Co., Inc., 1956), pp. 201-207. Reprinted with permission.
*Illustrative educational objectives selected from the literature.

1.10 Knowledge of Specifics

The recall of specific and isolable bits of information. The emphasis is on symbols with concrete referents. This material, which is at a very low level of abstraction, may be thought of as the elements from which more complex and abstract forms of knowledge are built.

1.11 Knowledge of Terminology

Knowledge of the referents for specific symbols (verbal and non-verbal). This may include knowledge of the most generally accepted symbol referent, knowledge of the variety of symbols which may be used for a single referent, or knowledge of the referent most appropriate to a given use of a symbol.

*To define technical terms by giving their attributes, properties, or relations.

*Familiarity with a large number of words in their common range of meanings.

1.12 Knowledge of Specific Facts

Knowledge of dates, events, persons, places, etc. This may include very precise and specific information such as the specific date or exact magnitude of a phenomenon. It may also include approximate or relative information such as an approximate time period or the general order of magnitude of a phenomenon.

*The recall of major facts about particular cultures.

*The possession of a minimum knowledge about the organisms studied in the laboratory.

1.20 Knowledge of Ways and Means of Dealing With Specifics

Knowledge of the ways of organizing, studying, judging, and criticizing. This includes the methods of inquiry, the chronological sequences, and the standards of judgment within a field as well as the patterns of organization through which the areas of the fields themselves are determined and internally organized. This knowledge is at an intermediate level of abstraction between specific knowledge on the one hand and knowledge of universals on the other. It does not so much demand the activity of the student in using the materials as it does a more passive awareness of their nature.

1.21 Knowledge of Conventions

Knowledge of characteristic ways of treating and presenting ideas and phenomena. For purposes of communication and consistency, workers in a field employ usages, styles, practices, and forms which best suit their purposes and/or which appear to suit best the phenomena with which they deal. It should be recognized that although these forms and conventions are likely to be set up on arbitrary, accidental, or authoritative bases, they are retained because of the general agreement or concurrence of individuals concerned with the subject, phenomena, or problem.

*Familiarity with the forms and conventions of the major types of works, e.g., verse, plays, scientific papers, etc.

*To make pupils conscious of correct form and usage in speech and writing.

1.22 Knowledge of Trends and Sequences

Knowledge of the processes, directions, and movements of phenomena with respect to time.

*Understanding of the continuity and development of American culture as exemplified in American life.

*Knowledge of the basic trends underlying the development of public assistance programs.

1.23 Knowledge of Classifications and Categories

Knowledge of the classes, sets, divisions, and arrangements which are regarded as fundamental for a given subject field, purpose, argument, or problem.

*To recognize the area encompassed by various kinds of problems or materials.

*Becoming familiar with a range of types of literature.

1.24 Knowledge of Criteria

Knowledge of the criteria by which facts, principles, opinions, and conduct are tested or judged.

*Familiarity with criteria for judgment appropriate to the type of work and the purpose for which it is read.

*Knowledge of criteria for the evaluation of recreational activities.

1.25 Knowledge of Methodology

Knowledge of the methods of inquiry, techniques, and procedures employed in a particular subject field as well as those employed in investigating particular problems and phenomena. The emphasis here is on the individual's knowledge of the method rather than his ability to use the method.

*Knowledge of scientific methods for evaluating health concepts.

*The student shall know the methods of attack relevant to the kinds of problems of concern to the social sciences.

1.30 Knowledge of the Universals and Abstractions in a Field

Knowledge of the major schemes and patterns by which phenomena and ideas are organized. These are the large structures, theories, and generalizations which dominate a subject field or which are quite generally used in studying phenomena or solving problems. These are at the highest levels of abstraction and complexity.

1.31 Knowledge of Principles and Generalizations

Knowledge of particular abstractions which summarize observations of phenomena. These are the abstractions which are of value in explaining, describing, predicting, or in determining the most appropriate and relevant action or direction to be taken.

*Knowledge of the important principles by which our experience with biological phenomena is summarized.

*The recall of major generalizations about particular cultures.

1.32 Knowledge of Theories and Structures

Knowledge of the body of principles and generalizations together with their interrelations which present a clear, rounded, and systematic view of a complex phenomenon, problem, or field. These are the most abstract formulations, and they can be used to show the interrelation and organization of a great range of specifics.

*The recall of major theories about particular cultures.

*Knowledge of a relatively complete formulation of the theory of evolution.

Intellectual Abilities and Skills

Abilities and skills refer to organized modes of operation and generalized techniques for dealing with materials and problems. The materials and problems may be of such a nature that little or no specialized and technical information is required. Such information as is required can be assumed to be part of the individual's general fund of knowledge. Other problems may require specialized and technical information at a rather high level such that specific knowledge and skill in dealing with the problem and the materials are required. The abilities and skills objectives emphasize the mental processes of organizing and reorganizing material to achieve a particular purpose. The materials may be given or remembered.

2.00 Comprehension

This represents the lowest level of understanding. It refers to a type of understanding or apprehension such that the individual knows what is being communicated and can make use of the material or idea being communicated without necessarily relating it to other material or seeing its fullest implications.

2.10 Translation

Comprehension as evidenced by the care and accuracy with which the communication is paraphrased or rendered from one language or form of communication to another. Translation is judged on the basis of faithfulness and accuracy, that is, on the extent to which the material in the original communication is preserved although the form of the communication has been altered.

*The ability to understand non-literal statements (metaphor, symbolism, irony, exaggeration).

*Skill in translating mathematical verbal material into symbolic statements and vice versa.

2.20 Interpretation

The explanation or summarization of a communication. Whereas translation involves an objective part-for-part rendering of a communication, interpretation involves a reordering, rearrangement, or a new view of the material.

*The ability to grasp the thought of the work as a whole at any desired level of generality.

*The ability to interpret various types of social data.

2.30 Extrapolation

The extension of trends or tendencies beyond the given data to determine implications, consequences, corollaries, effects, etc., which are in accordance with the conditions described in the original communication.

*The ability to deal with the conclusions of a work in terms of the immediate inference made from the explicit statements.

*Skill in predicting continuation of trends.

3.00 Application

The use of abstractions in particular and concrete situations. The abstractions may be in the form of general ideas, rules of procedures, or generalized methods. The abstractions may also be technical principles, ideas, and theories which must be remembered and applied.

*Application to the phenomena discussed in one paper of the scientific terms or concepts used in other papers.

*The ability to predict the probable effect of a change in a factor on a biological situation previously at equilibrium.

4.00 Analysis

The breakdown of a communication into its constituent elements or parts such that the relative hierarchy of ideas is made clear and/or the relations between the ideas expressed are made explicit. Such analyses are intended to clarify the communication, to indicate how the communication is organized, and the way in which it manages to convey its effects, as well as its basis and arrangement.

4.10 Analysis of Elements

Identification of the elements included in a communication.

*The ability to recognize unstated assumptions.

*Skill in distinguishing facts from hypotheses.

4.20 Analyses of Relationships

The connections and interactions between elements and parts of a communication.

*Ability to check the consistency of hypotheses with given information and assumptions.

*Skill in comprehending the interrelationships among the ideas in a passage.

4.30 Analysis of Organizational Principles

The organization, systematic arrangement, and structure which hold the communication together. This includes the "explicit" as well as "implicit" structure. It includes the bases, necessary arrangement, and the mechanics which make the communication a unit.

*The ability to recognize form and pattern in literary or artistic works as a means of understanding their meaning.

*Ability to recognize the general techniques used in persuasive materials, such as advertising, propaganda, etc.

5.00 Synthesis

The putting together of elements and parts so as to form a whole. This involves the process of working with pieces, parts, elements, etc., and arranging and combining them in such a way as to constitute a pattern or structure not clearly there before.

5.10 Production of a Unique Communication

The development of a communication in which the writer or speaker attempts to convey ideas, feelings, and/or experiences to others.

*Skill in writing, using an excellent organization of ideas and statements.

*Ability to tell a personal experience effectively.

5.20 Production of a Plan, or Proposed Set of Operations

The development of a plan of work or the proposal of a plan of operations. The plan should satisfy requirements of the task which may be given to the student or which he may develop for himself.

*Ability to propose ways of testing hypotheses.

*Ability to plan a unit of instruction for a particular teaching situation.

5.30 Derivation of a Set of Abstract Relations

The development of a set of abstract relations either to classify or explain particular data or phenomena, or the deduction of propositions and relations from a set of basic propositions or symbolic representations.

*Ability to formulate appropriate hypotheses based upon an analysis of factors involved, and to modify such hypotheses in the light of new factors and considerations.

*Ability to make mathematical discoveries and generalizations.

6.00 Evaluation

Judgments about the value of material and methods for given purposes. Quantitative and qualitative judgments about the extent to which material and methods satisfy criteria. Use of a standard of appraisal. The criteria may be those determined by the student or those which are given to him.

6.10 Judgments in Terms of Internal Evidence

Evaluation of the accuracy of a communication from such evidence as logical accuracy, consistency, and other internal criteria.

*Judging by internal standards, the ability to assess general probability of accuracy in reporting facts from the care given to exactness of statement, documentation, proof, etc.

*The ability to indicate logical fallacies in arguments.

6.20 Judgments in Terms of External Criteria

Evaluation of material with reference to selected or remembered criteria.

*The comparison of major theories, generalizations, and facts about particular cultures.

*Judging by external standards, the ability to compare a work with the highest known standards in its field—especially with other works of recognized excellence.

Exercises: Taxonomy of Educational Objectives: Cognitive Domain

1. Suppose you have just swallowed the notions in Bloom's Taxonomy, hook, line, and sinker. How will you now behave as a teacher? Give examples.
2. Choose a textbook you expect to teach from and pick a specific set of exercises. Classify each exercise using the categories of the taxonomy. Careful students usually find little that goes beyond the knowledge and comprehension levels. Did that happen to you.? Explain.
3. Of the six levels given in the taxonomy, which do you find most important? Why?
4. How does the problem of test writing relate to the taxonomy? Are some levels easier to measure than others? Explain.
5. With what levels do you feel Bruner would be most concerned? Why?
6. What dangers do you see in a teacher using a taxonomy such as this one? Explain. What advantages do you see? Explain.
7. Can the instructional weaknesses given by Brownell in "The Progressive Nature of Learning in Mathematics" be weaknesses for the use of Bloom's Taxonomy? Illustrate.
8. Consider the topic of adding positive and negative integers. Give an example of an objective for this topic for each of the levels: knowledge, comprehension, application, analysis, synthesis, and evaluation.
9. What do you think of the order of the six levels? Are they ordered the way you would order them? Is evaluation really the highest order? Explain.
10. Compare the Bloom, et al., Taxonomy of Cognitive Objectives with Gagné's eight types of learning.

Condensed Version of the Taxonomy of Educational Objectives: Affective Domain

Introduction to Taxonomy of Educational Objectives: Affective Domain

With a great deal of hesitation and nearly eight years after the cognitive taxonomy, the promised taxonomy for the affective domain was published. The authors had severe misgivings about the material and consented to its publication because of the urging of others. If writing a taxonomy for the cognitive domain was difficult, writing a taxonomy to deal with attitudes (affective domain) was thought to be near impossible. Here is a condensed version of the attempt. It is not the classic in education today that the cognitive taxonomy is. However, read this taxonomy as you read the taxonomy for the cognitive domain. What does it mean for you in your classroom.

1.0 Receiving (Attending)

At this level we are concerned that the learner be sensitized to the existence of certain phenomena and stimuli; that is, that he be willing to receive or to attend to them. This is clearly the first and crucial step if the learner is to be properly oriented to learn what the teacher intends that he will. To indicate that this is the bottom rung of the ladder, however, is not at all to imply that the teacher is starting *de novo*. Because of previous experience (formal or informal), the student brings to each situation a point of view or set which may facilitate or hinder his recognition of the phenomena to which the teacher is trying to sensitize him.

The category of *Receiving* has been divided into three subcategories to indicate three different levels of attending to phenomena. While the division points between the subcategories are arbitrary, the subcategories do represent a continuum. From an extremely passive position or role on the part of the learner, where the sole responsibility for the evocation of the behavior rests with the teacher—that is, the responsibility rests with him for "capturing" the student's attention—the continuum extends to a point at which the learner directs his attention, at least at a semiconscious level, toward the preferred stimuli.

David R. Krathwohl, Benjamin S. Bloom, and Betram S. Masia, "Condensed Version of the Taxonomy of Educational Objectives," *Taxonomy of Educational Objectives, Handbook II: Affective Domain* (New York: David McKay Co., Inc., 1964), pp. 176-185. Reprinted with permission.

1.1 Awareness

Awareness is almost a cognitive behavior. But unlike *Knowledge,* the lowest level of the cognitive domain, we are not so much concerned with a memory of, or ability to recall, an item or fact as we are that, given appropriate opportunity, the learner will merely be conscious of something—that he take into account a situation, phenomenon, object, or stage of affairs. Like *Knowledge* it does not imply an assessment of the qualities or nature of the stimulus, but unlike *Knowledge* it does not necessarily imply attention. There can be simple awareness without specific discrimination or recognition of the objective characteristics of the object, even though these characteristics must be deemed to have an effect. The individual may not be able to verbalize the aspects of the stimulus which cause the awareness.

> Develops awareness of aesthetic factors in dress, furnishings, architecture, city design, good art, and the like.

> Develops some consciousness of color, form, arrangement, and design in the objects and structures around him and in descriptive or symbolic representations of people, things, and situations.[1]

1.2 Willingness to Receive

In this category we have come a step up the ladder but are still dealing with what appears to be cognitive behavior. At a minimum level, we are here describing the behavior of being willing to tolerate a given stimulus, not to avoid it. Like *Awareness,* it involves a neutrality or suspended judgment toward the stimulus. At this level of the continuum the teacher is not concerned that the student seek it out, nor even, perhaps, that in an environment crowded with many other stimuli the learner will necessarily attend to the stimulus. Rather, at worst, given the opportunity to attend in a field with relatively few competing stimuli, the learner is not actively seeking to avoid it. At best, he is willing to take notice of the phenomenon and give it his attention.

> Attends (carefully) when others speak—in direct conversation, on the telephone, in audiences.

> Appreciation (tolerance) of cultural patterns exhibited by individuals from other groups—religious, social, political, economic, national, etc.

> Increase in sensitivity to human need and pressing social problems.

1.3 Controlled or Selected Attention

At a somewhat higher level we are concerned with a new phenomenon, the differentiation of a given stimulus into figure and ground at a conscious or perhaps semiconscious level—the differentiation of aspects of a stimulus which is perceived as clearly marked off from adjacent impressions. The perception is still without tension or assessment, and the student may not know the technical terms or symbols with which to describe it correctly or precisely to others. In some instances it may refer not so much to the selectivity of attention as to the control of attention, so that when certain stimuli are present they will be attended to. There is an

---

[1]Illustrative objectives selected from the literature follow the description of each subcategory.

element of the learner's controlling the attention here, so that the favored stimulus is selected and attended to despite competing and distracting stimuli.

> Listens to music with some discrimination as to its mood and meaning and with some recognition of the contributions of various musical elements and instruments to the total effect.

> Alertness toward human values and judgments on life as they are recorded in literature.

2.0 Responding

At this level we are concerned with responses which go beyond merely attending to the phenomenon. The student is sufficiently motivated that he is not just 1.2 *Willing to attend,* but perhaps it is correct to say that he is actively attending. As a first stage in a "learning by doing" process the student is committing himself in some small measure to the phenomena involved. This is a very low level of commitment, and we would not say at this level that this was "a value of his" or that he had "such and such an attitude." These terms belong to the next higher level that we describe. But we could say that he is doing something with or about the phenomenon besides merely perceiving it, as would be true at the next level below this of 1.3 *Controlled or selected attention.*

This is the category that many teachers will find best describes their "interest" objectives. Most commonly we use the term to indicate the desire that a child become sufficiently involved in or committed to a subject, phenomenon, or activity that he will seek it out and gain satisfaction from working with it or engaging in it.

2.1 Acquiescence in Responding

We might use the word "obedience" or "compliance" to describe this behavior. As both of these terms indicate, there is a passiveness so far as the initiation of the behavior is concerned, and the stimulus calling for this behavior is not subtle. Compliance is perhaps a better term than obedience, since there is more of the element of reaction to a suggestion and less of the implication of resistance or yielding unwillingly. The student makes the response, but he has not fully accepted the necessity for doing so.

> Willingness to comply with health regulations.

> Obeys the playground regulations.

2.2 Willingness to Respond

The key to this level is in the term "willingness," with its implication of capacity for voluntary activity. There is the implication that the learner is sufficiently committed to exhibiting the behavior that he does so not just because of a fear of punishment, but "on his own" or voluntarily. It may help to note that the element of resistance or of yielding unwillingly, which is possibly present at the previous level, is here replaced with consent or proceeding from one's own choice.

> Acquaints himself with significant current issues in international, political, social, and economic affairs through voluntarry reading and discussion.

Acceptance of responsibility for his own health and for the protection of the health of others.

2.3 Satisfaction in Response

The additional element in the step beyond the *Willingness to respond* level, the consent, the assent to responding, or the voluntary response, is that the behavior is accompanied by a feeling of satisfaction, an emotional response, generally of pleasure, zest, or enjoyment. The location of this category in the hierarchy has given us a great deal of difficulty. Just where in the process of internalization the attachment of an emotional response, kick, or thrill to a behavior occurs has been hard to determine. For that matter there is some uncertainty as to whether the level of internalization at which it occurs may not depend on the particular behavior. We have even questioned whether it should be a category. If our structure is to be a hierarchy, then each category should include the behavior in the next level below it. The emotional component appears gradually through the range of internalization categories. The attempt to specify a given position in the hierarchy as *the* one at which the emotional component is added is doomed to failure.

The category is arbitrarily placed at this point in the hierarchy where it seems to appear most frequently and where it is cited as or appears to be an important component of the objectives at this level on the continuum. The category's inclusion at this point serves the pragmatic purpose of reminding us of the presence of the emotional component and its value in the building of affective behaviors. But it should not be thought of as appearing and occurring at this one point in the continuum and thus destroying the hierarchy which we are attempting to build.

Enjoyment of self-expression in music and in arts and crafts as another means of personal enrichment.

Finds pleasure in reading for recreation.

Takes pleasure in conversing with many different kinds of people.

3.0 Valuing

This is the only category headed by a term which is in common use in the expression of objectives by teachers. Further, it is employed in its usual sense: that a thing, phenomenon, or behavior has worth. This abstract concept of worth is in part a result of the individual's own valuing or assessment, but it is much more a social product that has been slowly internalized or accepted and has come to be used by the student as his own criterion of worth.

Behavior categorized at this level is sufficiently consistent and stable to have taken on the characteristics of a belief or an attitude. The learner displays this behavior with sufficient consistency in appropriate situations that he comes to be perceived as holding a value. At this level, we are not concerned with the relationships among values but rather with the internalization of a set of specified, ideal, values. Viewed from another standpoint, the objectives classified here are the prime stuff from which the conscience of the individual is developed into active control of behavior.

This category will be found appropriate for many objectives that use the term "attitude" (as well as, of course, "value").

An important element of behavior characterized by *Valuing* is that it is motivated, not by the desire to comply or obey, but by the individual's commitment to the underlying value guiding the behavior.

3.1 Acceptance of a Value

At this level we are concerned with the ascribing of worth to a phenomenon, behavior, object, etc. The term "belief," which is defined as "the emotional acceptance of a proposition or doctrine upon what one implicitly considers adequate ground" (English and English, 1958, p. 64), describes quite well what may be thought of as the dominant characteristic here. Beliefs have varying degrees of certitude. At this lowest level of *Valuing* we are concerned with the lowest levels of certainty; that is, there is more of a readiness to re-evaluate one's position than at the higher levels. It is a position that is somewhat tentative.

One of the distinguishing characteristics of this behavior is consistency of response to the class of objects, phenomena, etc. with which the belief or attitude is identified. It is consistent enough so that the person is perceived by others as holding the belief or value. At the level we are describing here, he is both sufficiently consistent that others can identify the value, and sufficiently committed that he is willing to be so identified.

Continuing desire to develop the ability to speak and write effectively.

Grows in his sense of kinship with human beings of all nations.

3.2 Preference for a Value

The provision for this subdivision arose out of a feeling that there were objectives that expressed a level of internalization between the mere acceptance of a value and commitment or conviction in the usual connotation of deep involvement in an area. Behavior at this level implies not just the acceptance of a value to the point of being willing to be identified with it, but the individual is sufficiently committed to the value to pursue it, to seek it out, to want it.

Assumes responsibility for drawing reticent members of a group into conversation.

Deliberately examines a variety of viewpoints on controversial issues with a view to forming opinions about them.

Actively participates in arranging for the showing of contemporary artistic efforts.

3.3 Commitment

Belief at this level involves a high degree of certainty. The ideas of "conviction" and "certainty beyond a shadow of a doubt" help to convey further the level of behavior intended. In some instances this may border on faith, in the sense of it being a firm emotional acceptance of a belief upon admittedly nonrational grounds. Loyalty to a position, group, or cause would also be classified here.

The person who displays behavior at this level is clearly perceived as holding the value. He acts to further the thing valued in some way, to extend the possibility of his developing it, to deepen his involvement with it and with the things representing it. He tries to convince others and seeks converts to his cause. There is a tension here which needs to be satisfied; action is the result of an aroused need or drive. There is a real motivation to act out the behavior.

> Devotion to those ideas and ideals which are the foundations of democracy.

> Faith in the power of reason and in methods of experiment and discussion.

4.0 Organization

As the learner successively internalizes values, he encounters situations for which more than one value is relevant. Thus necessity arises for (*a*) the organization of the values into a system, (*b*) the determination of the interrelationships among them, and (*c*) the establishment of the dominant and pervasive ones. Such a system is built gradually, subject to change as new values are incorporated. This category is intended as the proper classification for objectives which describe the beginnings of the building of a value system. It is subdivided into two levels, since a prerequisite to interrelating is the conceptualization of the value in a form which permits organization. *Conceptualization* forms the first subdivision in the organization process, *Organization of a value system* the second.

While the order of the two subcategories seems appropriate enough with reference to one another, it is not so certain that 4.1 *Conceptualization of a value* is properly placed as the next level above 3.3 *Commitment.* Conceptualization undoubtedly begins at an earlier level for some objectives. Like 2.3 *Satisfaction in response,* it is doubtful that a single completely satisfactory location for this category can be found. Positioning it before 4.2 *Organization of a value system* appropriately indicates a prerequisite of such a system. It also calls attention to a component of affective growth that occurs at least by this point on the continuum but may begin earlier.

4.1 Conceptualization of a Value

In the previous category, 3.0 *Valuing,* we noted that consistency and stability are integral characteristics of the particular value or belief. At this level (4.1) the quality of abstraction or conceptualization is added. This permits the individual to see how the value relates to those that he already holds or to new ones that he is coming to hold.

Conceptualization will be abstract, and in this sense it will be symbolic. But the symbols need not be verbal symbols. Whether conceptualization first appears at this point on the affective continuum is a moot point, as noted above.

> Attempts to identify the characteristics of an art object which he admires.

> Forms judgments as to the responsibility of society for conserving human and material resources.

4.2 Organization of a Value System

Objectives properly classified here are those which require the learner to bring together a complex of values, possibly disparate values, and to bring these into an ordered relationship with one another. Ideally, the ordered relationship will be one which is harmonious and internally consistent. This is, of course, the goal of such objectives, which seek to have the student formulate a philosophy of life. In actuality, the integration may be something less than entirely harmonious. More likely the relationship is better described as a kind of dynamic equilibrium which is, in part, dependent upon those portions of the environment which are salient at any point in time. In many instances the organization of values may result in their synthesis into a new value or value complex of a higher order.

> Weighs alternative social policies and practices against the standards of the public welfare rather than the advantage of specialized and narrow interest groups.

> Develops a plan for regulating his rest in accordance with the demands of his activities.

5.0 Characterization by a Value or Value Complex

At this level of internalization the values already have a place in the individual's value hierarchy, are organized into some kind of internally consistent system, have controlled the behavior of the individual for a sufficient time that he has adapted to behaving this way; and an evocation of the behavior no longer arouses emotion or affect except when the individual is threatened or challenged.

The individual acts consistently in accordance with the values he has internalized at this level, and our concern is to indicate two things: (*a*) the generalization of this control to so much of the individual's behavior that he is described and characterized as a person by these pervasive controlling tendencies, and (*b*) the integration of these beliefs, ideas, and attitudes into a total philosophy or world view. These two aspects constitute the subcategories.

5.1 Generalized Set

The generalized set is that which gives an internal consistency to the system of attitudes and values at any particular moment. It is selective responding at a very high level. It is sometimes spoken of as a determining tendency, an orientation toward phenomena, or a predisposition to act in a certain way. The generalized set is a response to highly generalized phenomena. It is a persistent and consistent response to a family of related situations or objects. It may often be an unconscious set which guides action without conscious forethought. The generalized set may be thought of as closely related to the idea of an attitude cluster, where the commonality is based on behavioral characteristics rather than the subject or object of the attitude. A generalized set is a basic orientation which enables the individual to reduce and order the complex world about him and to act consistently and effectively in it.

> Readiness to revise judgments and to change behavior in the light of evidence.

> Judges problems and issues in terms of situations, issues, purposes, and consequences involved rather than in terms of fixed, dogmatic precepts or emotionally wishful thinking.

5.2 Characterization

This, the peak of the internalization process, includes those objectives which are broadest with respect both to the phenomena covered and to the range of behavior which they comprise. Thus, here are found those objectives which concern one's view of the universe, one's philosophy of life, one's *Weltanschauung*—a value system having as its object the whole of what is known or knowable.

Objectives categorized here are more than generalized sets in the sense that they involve a greater inclusiveness and, within the group of attitudes, behaviors, beliefs, or ideas, an emphasis on internal consistency. Though this internal consistency may not always be exhibited behaviorally by the students toward whom the objective is directed, since we are categorizing teachers' objectives, this consistency feature will always be a component of *Characterization* objectives.

As the title of the category implies, these objectives are so encompassing that they tend to characterize the individual almost completely.

Develops for regulation of one's personal and civic life a code of behavior based on ethical principles consistent with democratic ideals.

Develops a consistent philosophy of life.

Exercises: Taxonomy of Educational Objectives: Affective Domain

1. Perhaps you noticed the lack of examples for each level in the taxonomy. Using mathematics, give examples of each of the thirteen subcategories identified in the taxonomy.

2. What do you see as the strengths and weaknesses of this taxonomy? Explain.

3. How would you measure students' attitude toward mathematics? If students have a desirable attitude, how will they behave in your class? How will they behave twenty years from now?

4. How do you react to the ordering of the levels—Receiving, Responding, Valuing, Organization, Characterization by a Value or a Value Complex? Explain.

5. Which of the five categories do you feel is most important? Why?

6. How can learning theory be related to taxonomy? Give examples.

Meaning and Skill—
Maintaining
the Balance

William A. Brownell

The subject given me poses a question. It is no academic question arising out of purely theoretical considerations. It is assumed that both meaning and computational competence are proper ends of instruction in arithmetic. It is implied that somehow or other both ends are not always achieved and that there is evidence that this is so.

Indeed there is such evidence. More than one school system has embarked upon a program of so-called meaningful arithmetic, only to discover that on standardized tests of computation and "problem solving" pupils do none too well. In such schools officials and teachers are likely to believe that they have made a bad bargain; and school patrons are likely to support them vehemently in this belief.

We may try to convince all concerned that the instruments used to evaluate learning are inappropriate, or at least imperfect and incomplete. True, standardized tests rarely if ever provide means to assess understanding of arithmetical ideas and procedures. Hence, the program of meaningful instruction, even if well managed, has no chance to reveal directly and explicitly its contribution to this aspect of learning. On the other hand, can we deny that the learning outcomes that *are* measured are of no significance? To do so is to say in effect the computational skill is of negligible importance, and we can hardly justify this position.

Why is there now the necessity to talk about establishing and maintaining the desirable kind of balance between meaning on the one hand and computational competence on the other?

Sources of the Dilemma

1. Incomplete Exposition of "Meaningful Arithmetic"

Perhaps those of us who have advocated meaningful learning in arithmetic are at fault. In objecting to the drill conception of the subject prevalent not so long ago, we may have failed to point out that practice for proficiency in skills has its place, too. It is questionable whether any who have spoken for meaningful instruction ever proposed that children be allowed to leave our schools unable to compute accurately, quickly, and confidently. I am sure that all, if asked, would have rejected this notion completely. But we may not have *said* so, or said so often enough or vigorously enough. Our comparative silence on this score may easily have been misconstrued to imply indifference about proficiency in computation.

William A. Brownell, "Meaning and Skill—Maintaining the Balance." Reprinted from the *Arithmetic Teacher* 15 (October 1956): 129-136. ©1956 by the National Council of Teachers of Mathematics. Reprinted with permission.

Paper read before the Elementary Section of the N.C.T.M. at the Milwaukee meeting of April 14, 1956.

If this has actually happened, we can scarcely blame classroom teachers if they have neglected computational skill as a learning outcome. It is characteristic of educational movements to behave pendulum-wise. When we correct, we tend to over-correct. Just this sort of thing seems to have happened in the teaching of arithmetic. In fleeing from over-reliance on one kind of practice, we may have fled too far. It is a curious state of affairs that those of us who deplored the limitations of this kind of practice must now speak out in its behalf and stress its positive usefulness.

2. Misunderstood Learning Theory

A second possible explanation for our dilemma may be found, as is frequently the case, in the practical business of education, in misinterpretations or misapplications of psychological theories of learning. Over-simplification of certain generalizations in learning theories as widely apart as are those of conditioning and of field theory could lead, and may have led, to the lessening of emphasis on practice in arithmetic.

a. Conditioning theory. According to the learning theory of one influential exponent of conditioning, once a response has been made to a stimulus a connection has been established. Thus, the child who says "Seven" in replying to the question, "How many are two and five?", sets up a connection between 2 and 5 as stimulus and 7 as response. If this is so—if the connection is actually formed by the one response—then at first glance practice might seem to be utterly purposeless.

But this inference is quite unfounded, as the psychologist in question makes abundantly clear. In saying that a connection is "established" he means no more than that a new neuromuscular pattern is available. He does *not* mean that our hypothetical child after the one experience will always, only, and instantly respond with "seven" when asked the sum of 2 and 5. If the response "seven" is to be the invariable one, moreover if the response is to be made in situations differing ever so slightly from the original one, then practice is required. In this theory of learning as conditioning, therefore, there is no comfort for those who would abandon practice as a means to promote the learning of arithmetical facts and skills.

b. Field theory. No more comfort is to be found in field theory of learning. It is often said that one experience of "insight" or "hindsight"—before, during, or after success—is enough; but enough for *what?* It may be all that is needed to understand a situation and the method of dealing with it. Yet, it is one thing to know the general rationale for solving a complicated mechanical puzzle but quite another thing to be able to manipulate the parts correctly with facility, ease, and speed.

So in arithmetic it is one thing to comprehend the mathematical principles governing decomposition in subtraction—something that can come from a single insightful experience—but another thing to be able to subtract quickly and correctly. Indeed if in examples like 73 - 47 and 52 - 19 a child who possesses this understanding always thinks through the complete logical explanation, his performance will be impaired, at least from the standpoint of speed. Understanding and skill are not identical, A single instance of insight may lead to understanding but will hardly produce skill. For skill, practice is necessary.

3. Influence of General Educational Theory

A third explanation for failure to stress computational competence is to be found in the recent history of educational theory. It must be remembered that the place of meaning and understanding in arithmetic has been generally recognized for not more than fifteen or twenty years. In 1935 we were still under the influence of somewhat sentimental and unrealistic notions both about children and about the course of their development. In extreme form these notions led to a kind of teaching that was anything but systematic. Indeed, what children learned, when they learned it, and how they learned it was left pretty much to the children themselves. Attempts to guide and direct learning and to organize learning experiences were frowned upon as "violating child nature" and as almost certainly productive of serious derangements of child personality. To those who held these views practice was anathema.

In the public schools, as contrasted with college departments of Education, this conception of the processes of learning and teaching did not gain much of a foothold. Nevertheless, it was in this climate of thought that stress on meaning in arithmetic put in its appearance. Those who were committed to the educational theory I have mentioned welcomed the new emphasis as confirming both what they did and what they did not do. If, deliberately or unwittingly, they accepted only the part of the emerging view most congenial to them, they committed an error that is very common and that is altogether human. Be that as it may, the consequences were none too good. True, these teachers may have been largely responsible for the quick and general endorsement of one aspect of meaningful arithmetic—learning with understanding. On the other hand, they could not themselves absorb the whole of it, and they remained hostile to practice in learning. Pupils taught by such teachers cannot be expected to make high scores on standardized arithmetic tests of skill in computation and "problem solving."

4. Inadequate Instruction on Meanings

I have suggested three hypotheses as explaining why we may not be obtaining balance between understanding and computational competence in arithmetic, on the assumption that our shortcomings relate to the latter (computational competence) rather than to the former. These three are: the possible failure of advocates of meaningful arithmetic to emphasize sufficiently the importance of practice in acquiring arithmetical skills; misinterpretations of psychological theories of learning which have had the effect of minimizing the place of practice; and the unwillingness of some teachers, who believe completely that arithmetic must be made intelligible to children, to provide the practice necessary for computational proficiency. May I add a fourth? It is that we may not as yet be doing a very good job in teaching arithmetical meanings as they should be taught.

There is ample evidence in psychological research on learning that the effects of understanding are cumulative. There is also ample evidence, if not in arithmetic, then in other types of learning, that the greater the degree of understanding, the less the amount of practice necessary to promote and to fix learning. If these truths are sound—and I think they are—then they should hold in the field of arithmetical learning. It follows that computational skills among school children would be greater than they are if we *really* taught them to understand what they learn.

Again I remind you that meaningful arithmetic, as this phrase is commonly used, is a newcomer in educational thought. Many teachers, trained in instructional procedures suitable, say, to a view of arithmetic as a tool or a drill subject, find it difficult to comprehend fully what meaningful arithmetic is and what it implies for the direction of learning. Others than myself, I am sure, have, in conferences with teachers, been somewhat surprised to note that some of them are unfamiliar with major ideas in this conception and with methods of instruction adapted thereto.

Perhaps the commonest instructional error is, in a different context, the same one that has always distorted learning in arithmetic, namely, the acceptance of memorized responses in place of insistence upon understanding. Mathematical relationships, principles, and generalizations are couched in language. For example, the relationship between a given set of addends and their sum is expressed verbally in some some such way as: "The order of the numbers to be added does not change the sum." It is about as easy for a child to master this statement by rote memorization as to master the number fact, $8 - 7 = 1$, and the temptation is to be satisfied when children can repeat the words of the generalization *verbatim*. Similarly, the rationale of computation in examples such as: $33 + 48$ and $71 - 16$, makes use of concepts deriving from our number system and our notions of place value. But many a child glibly uses the language of "tens" and "ones" with no real comprehension of what he is saying. Such learning is a waste of time. To use an Irish bull, the meanings have no meaning.

I intend no criticism of teachers. Until recently there have been few professional books of high quality to set forth the mathematics of arithmetic and to describe the kind of instruction needed. Moreover, many teachers have had no access to these few books. Again, until recently not many courses of study and teachers' manuals for textbook series have been of much help. It is not strange, therefore, that though meaningful arithmetic is adopted in a given school system, not all members of the teaching staff are well equipped to teach it. As a result, their pupils, denied a full and intelligent treatment of arithmetic as a body of rational ideas and procedures, have been unable to bring to computation all the aid that could come through understanding.

Toward a Solution

So much for possible explanations—explanations of a general character—for our failure to keep meaning and skill in balance. I have no way of knowing the reality of any of the four hypotheses suggested or of the extent of its validity, to say nothing of the degree to which, taken together, the four account adequately for the situation. The fact remains that something needs to be done. What is the remedy?

Certainly we shall not get very far as long as we think of understanding and practice in absolute terms. I have deliberately done this so far in order to examine the issue in simple terms. Actually, it is erroneous to conceive of understanding as if it were either totally present or totally absent. Instead, there are degrees or levels of understanding. Likewise, not all forms of practice are alike. Rather, there are different types, and they have varying effects in learning.

Levels of Understanding

Consider the example 26 + 7. The child who first lays out twenty-six separate objects, next seven more objects, and then determines the total by counting the objects one by one has a meaning for the operation. So has the child who counts silently, starting with 26. So has the child who breaks the computation into two steps, 26 + 4 = 30 and 30 + 3 = 33. So has the child who employs the principle of adding by endings,—6 + 7 = 13, so 26 + 7 = 33. So has the child who, capable of all these types of procedure, nevertheless recognizes 33 at once as the sum of 26 and 7. All these children "understand," but their understandings may be said to represent different points in the learning curve. The counter is at the bottom, and the child who through understanding has habituated his response "thirty-three" so that it comes automatically is at the top of a series of progressively higher levels of performance.

All these levels of performance or of understanding are good, depending upon the stage of learning when they are used. For instance, finding the sum of 26 and 7 by counting objects is a perfectly proper way of meeting the demand at first; but it is not the kind of performance we want of a child in grade 4. At some time in that grade, or earlier, he should arrive at the stage when he can announce the sum correctly, quickly, and confidently, with a maximum of understanding.

It is a mistake to believe that this last stage can be achieved at once, by command as it were. When a child is required to perform at a level higher than he has achieved, he can do only one of three things. (a) He can refuse to learn, and his refusal may take the form; "I won't," or "I can't," or "I don't care," the last named signifying frustration and indifference which we should seek to prevent at all costs. (b) Or, he can acquire such proficiency at the level he *has* attained that he will be credited for thinking at the level desired. Many children develop such expertness in silent counting that, in the absence of close observation and questioning, they are believed to have procedures much beyond those they do have. (c) Or, third, he may try to do what the teacher seems to want. If an immediate answer is apparently expected, he will supply one, by guessing or by recalling a memorized answer devoid of meaning. If he guesses, obviously he makes no progress at all in learning. If he memorizes, only unremitting practice will keep the association alive; and if he forgets, he is helpless or must drop back to a very immature level of performance such as counting objects or marks by 1's.

For the stage of performance we should aim for ultimately, as in the case of the simple number facts, higher-decade facts, and computational skills, we have no standard term. We may use the word "memorization" to refer to what a child does when he learns to say "Four and two are six" without understanding much about the numbers involved, about the process of addition, or about the idea of equivalence. If we employ "memorization" in this sense, then that word is inappropriate for the last step in the kind of learning we should foster. Hence for myself I have adopted the phrase "meaningful habituation." "Habituation" describes the almost automatic way in which the required response is invariably made; "meaningful" implies that the seemingly simple behavior has a firm basis in understanding. The particular word or phrase for this last step in meaningful learning is unimportant; but the *idea,* and its difference from "memorization," *are* important.

Teaching meaningfully consists in directing learning in such a way that children ascend, as it were, a stairway of levels of thinking arithmetically to the level of meaningful habituation in those aspects of arithmetic which should be thoroughly mastered, among them the basic computational skills. Too many pupils, even some supposedly taught through understanding, do not reach this last stage. Instead, they stop short thereof; and even if they are intelligent about what they do when they compute, they acquire little real proficiency. In instances of this kind both learning and teaching have been incomplete.

How are teachers to know the status of their pupils with respect to progress toward meaningful habituation? Little accurate information is to be had from their written work, for both correct and incorrect answers can be obtained in many ways, and inference is dangerous. Insightful observation and pupils' oral reports volunteered or elicited through questioning are more fruitful sources of authentic data. Since children differ so much in their thought procedures, a good deal of this probing must be done individually. One of the most fruitful devices I can suggest for this probing consists in noting what children do in the presence of error.

I recall a conversation with a fourth grade girl whom I knew very well and who was having difficulty in learning—in her case, in memorizing—the multiplication facts. I asked her—her name was June—"How many are five times nine?" (This form of expression was used in her school instead of the better "How many are five nines?") Immediately she responded, "Forty-five." When I shook my head and said, "No, forty-six," she was clearly upset. Her reply, after some hesitation, was, "No, it's forty-five." When I insisted that the correct product is 46, June said, "Well, that isn't the way I learned it." I suggested that perhaps she had learned the wrong answer. Her next statement was, "Well, that's what my teacher told me." This time I told her that she may have misunderstood her teacher or that her teacher was wrong. June was obviously puzzled; then she resorted to whispering the table, "One times nine is nine; two times nine is eighteen," and so on, until she reached "five times nine is forty-five." Again I shook my head and said, "Forty-six." When she was unable to reconcile my product with what she had become accustomed to say, I asked her, "June, have you no way of finding out whether forty-five or forty-six is the correct answer?" Her response was, "No, I just learned it as forty-five." Of course I did not leave her in her state of confusion; but the point of the illustration is, I hope, quite apparent: Her inability to deal with error was convincing evidence of the superficiality of her "learning" and of its worthlessness.

Compare my conversation with June with that I had with Anne, another fourth grade girl whom also I knew well and who, like the first girl, was learning the multiplication facts. When I asked Anne, "How many are five times nine?", the correct answer came at once, just as in June's case; but from here on, mark the difference. I introduced the error, saying that the product is 46, not 45. Anne looked at me in disgust and said, "Are you kidding?" I maintained my position that $5 \times 9 = 46$. Immediately she said, "Do you want me to prove it's forty-five?" I told her to go ahead if she thought she could. She answered, "Well, I can. Go to the blackboard." There I was instructed to write a column of five 9's and, not taking any chances with me, Anne told me to *count* the 9's to make sure I had 5. Next came the command, "Add them." When I deliberately made mistakes in addition, she corrected me, each time saying, "Do you want me to prove that, too?" Obviously, I had to arrive at a total of 45. Having done so, I said, "Oh, that's just a trick," to which she replied, "Do you want me to prove it another way?" Exposure to error held no terrors for this child; she did not become confused or fall back upon repetition of the multiplication table; nor did she cite her teacher as an authority. Instead, Anne had useful resources in the form of understandings that were quite lacking in the case of June whose discomfiture I have described.

Probing for understanding need not depend wholly on opportunities to work with individual children. On the contrary, there are possibilities also under conditions of group instruction when questions beginning with "How" and "Why" supplement the commoner questions starting with "What." The worth of valid knowledge concerning level of understanding is inestimable for the guidance of learning. The demands upon time are not inconsiderable, but no one should expect to get full knowledge concerning every pupil. The prospects are not hopeless if ingenuity is exercised and if the goal is set, not at 100% of knowledge, but more realistically at perhaps 20% more than is now ordinarily obtained.

Types of Practice

In crude terms, practice consists in doing the same thing over and over again. Actually, an individual never does the same thing twice, nor does he face the same situation twice, for the first reaction to a given situation alters both the organism and the situation. A changed being responds the second and the third time, and the situation is modified accordingly.

We must concede the truth of these facts. At the same time, for our purposes we may violate them a bit. Let us conceive of practice of whatever kind as falling somewhere along a continuum. At the one end of the continuum is practice in which the learner tries as best he can to repeat just what he has been doing. At the other end is practice in which the learner modifies his attack in dealing with what is objectively the same situation (or what to him are similar situations). We may call these extremes "repetitive practice" and "varied practice," respectively.

An instance of repetitive practice is memorizing the serial order of number names by rote or applying the series in the enumeration of groups of objects. An instance of varied practice is the attempt, by trying different approaches, to find steadily better ways of computing in such examples as $43 + 39$, $75 - 38$, $136 \div 4$, and 32×48. Between the two extreme types of practice are innumerable others, differing by degree in the extent to which either repetition or variation is employed. But again, for our purposes, we may disregard all the intervening sorts of practice: we could not possibly name them all, or describe them, or show their special contributions of learning.

Both repetitive and varied practice affect learning, but in quite unlike ways. For illustration we may choose an instance of learning outside of arithmetic, for example, the motor activity of swimming. Suppose the beginner engages in repetitive practice: What does he do, and what will happen? Well, he will continue to use as nearly as he can exactly the movements he employed the first time he was in deep water, and the result will be that he may become highly proficient in making just those movements. He will hardly become an expert swimmer, but he will become an expert in doing what he does, whether it be swimming or not.

On the other hand, suppose that the beginner engages in varied practice. In this case he will seek to *avoid* doing precisely what he did at the outset. He will discard uneconomical movements; he will try out other movements, select those that are most promising, and seek a final coordination that makes him a good swimmer. Then what will he do? He will change to repetitive practice, for, having the effective combination of movements he wants, he will seek to perfect it in order to become more proficient in it.

The differences between repetitive practice and varied practice, both in what the learner does and in what his practice produces, are clearly discernible when we think of motor activities. They are less easily identified when we think of ideational learning tasks like the number facts and computational skills. But the differences are there none the less. The child who counts and only counts in dealing with examples like 36 + 37 and 24 + 69 is employing repetitive practice. The more he counts, the more expert he becomes in counting; but the counting will not, and cannot, move him to a higher level of understanding and of performance. In contrast is the child who, through self-discovery or through instruction, tries different ways to add in such examples. Under guidance he can be led to adopt higher and higher levels of procedures until he is ready for meaningful habituation. If then he does not himself fix his automatic method of adding, he can be led to do so through repetitive practice. In any case it is safer to provide the repetitive practice in order to increase proficiency and make it permanent.

The distinction between repetitive and varied practice, in their nature and in their consequence, is not always recognized in teaching. If repetitive practice is introduced too soon, before understanding has been achieved, the result, for one thing, may be blind effort and frustration on the part of the learner. Or, it may fix his performance at a low level, the level he has attained. No new and better procedure can emerge from repetitive practice though it may appear under conditions of drill when a child, tired of repetition or disappointed in its result, abandons it in search of something new.

There may be an instructional error of another kind, one already alluded to. This error is to insist, to quote some, that "there is no place for drill in the modern conception of teaching." True, there is no place for unmotivated drill on ill-understood skills; but the statement goes too far in saying that there is no place at all for repetitive practice. How else, one may ask, is the final step of meaningful habituation to be made permanent; how else is real proficiency at this level of learning to be assured?

The kind of practice most beneficial at any time, then, is the kind best adapted to accomplish a given end. For illustration, let us return to June and Anne and their learning of the multiplication facts.

June was trying to master these facts by repetitive practice, by saying over and over and over again the special grouping of words for each separate fact. Her level of understanding of the numbers and relationships involved was close to zero. Unless engaged in almost ceaselessly, her repetition of verbalizations could give her little more than temporary control, and control, be it noted, of the verbalizations alone. Lapses of memory would be nearly fatal and would subject her to the hazards of guessing. In no way could her memorization of senseless phrases contribute much to sound learning of the facts themselves, not to mention the deficiences of its results for more advanced forms of computation and functional use. What June needed was not repetitive, but varied practice. By contrast, Anne, who was able to "prove" her announced products and who thereby demonstrated her full understanding of the relationship of the numbers, no longer needed varied practice and could safely and properly be encouraged to engage in repetitive practice.

We employ varied practice, then, if we wish the child to move upward from where he is toward meaningful habituation, and repetitive practice if we are endeavoring to produce true competence, economy, and permanence in this last stage of learning (or at any earlier stage which represents a type of performance of worth in itself). Practice has to be designed to fit the learner's needs, a fact which brings us back again to the individual and to the critical importance of accurate knowledge concerning his learning status.

In Conclusion

To sum up, the balance between meaning and skill has been upset, if indeed it ever was properly established. The reasons are many, some of them relating to educational theory in general, others to misconceptions of psychological theories of learning, others to failure to teach arithmetical meanings thoroughly, and still others to carry learning in the case of computation to the level I have denoted meaningful habituation, and then to fix learning at that level. I have discussed these matters at length, perhaps at unwarranted extent in view of the fact that the remedy for the situation can be stated briefly. The remedy I propose is as follows:

1. Accord to competence in computation its rightful place among the outcomes to be achieved through arithmetic;
2. Continue to teach essential arithmetical meanings, but make sure that these meanings are just that and that they contribute as they should to greater computational skill;
3. Base instruction on as complete data as are reasonably possible concerning the status of children as they progress toward meaningful habituation;
4. Hold repetitive practice to a minimum until this ultimate stage has been achieved; then provide it in sufficient amount to assure real mastery of skills, real competence in computing accurately, quickly, and confidently.

Exercises: Meaning and Skill—
Maintaining the Balance

1. Get a copy of a standardized test of computation and problem solving used by the schools in your area. Go through the test item by item and classify the items as primarily testing for meaning or skill. How many of each did you find? Does the test maintain the balance? What would be an ideal balance? Why?

2. Compare Brownell's position on misunderstood learning theory and the issue of practice with the position of Cronbach on practice. What kind of practice does each advocate? What does Dienes say about practice? State three different practical strategies which can be used to deal with practice in the classroom. What are their strengths? Weaknesses?

3. Brownell suggests that a single instance of insight may lead to understanding. Do you agree, or do you think one must practice the "understanding" too? Why? Justify your position by citing other readings.

4. Brownell states: "It follows that computational skills among school children would be greater than they are if we *really* taught them to understand what they learn." What evidence can you give to support this position? Examples? What evidence can you give to refute this position? Examples?

5. What is your reaction to Brownell's conversation with June? Can you give a similar example from your own experience? Is such learning worthless? Why? What about Anne? What difference does it make when one is an adult whether one is a June or an Anne? Explain. Give examples.

6. If one accepts the difference between repetitive practice and varied practice and uses them as Brownell recommends, what problems do individual differences cause? Suggest a possible solution.

7. Discuss the four remedies suggested by Brownell at the end of the article. Illustrate each with a specific example from the classroom.

8. Does the word *meaningful* have the same significance for Brownell and for Ausubel? Explain.

Discovery
in the Teaching
of Mathematics

Robert B. Davis

Whether one thinks of the aeronautical engineer versus the pilot, or of the pure mathematician versus the engineer, or of the economist versus the businessman, virtually every major field of human endeavor is split between a group of theorists and a group of practitioners. It seems to me that it is a sign of health when the two groups are able to communicate, and when they attempt to work closely together. Speaking as one of the clinical people or practitioners, I want to thank the theorists for their efforts at bridging the gap which sometimes seems to separate us.

My remarks are made from the practitioner's point of view, which is the only one I can legitimately claim. Based on the Madison Project work of the past eight years, I want to cite a few examples and a few ideas related to them.

Specifically, I have three goals: first, I want to give examples of what we have regarded as discovery experience for children, in the hope that some of you will be able to suggest a few of the ways in which these experiences differ from exposition, and even differ from one another. This last remark deserves emphasizing, for it is my present notion that there are many different kinds of discovery experience, and we confuse the issue badly when we treat discovery as a single well-defined kind of experience.

My second goal is to offer a few remarks attempting to interpret or describe some of the things we mean by discovery, and to explain why some of us believe in its importance.

Finally, I shall list a few things that might be called objectives of Madison Project teaching, for this, too, may clarify why we think we believe in "discovery."

Some Examples and Some Interpretations

The students in my first example were some low-IQ culturally deprived children in the seventh grade, some of whom were older than normal for grade seven. The topic was the matter of finding pairs of whole numbers that would satisfy linear equations of the form

$$(\square \times 3) + 2 = \triangle$$

Although we found that $\square, \triangle$ notation useful, and consequently did use it with these children, what I have written above could be translated into traditional x, y notation as

$$y = 3x + 2$$

Robert B. Davis, "Discovery in the Teaching of Mathematics," in Lee S. Shulman and Evan R. Keisler (eds.), *Learning by Discovery* (Chicago: Rand McNally, 1966), pp. 114-128. ©1966 Rand McNally & Company.

We passed out graph paper and suggested that suitable pairs of whole numbers be recorded according to the usual Cartesian use of coordinates. Very nearly all of the students made the obvious discovery that there is a simple linear pattern to the resulting dots. A considerable number of them actually applied this discovery, by extrapolating according to the pattern, then checking to see if their new points gave numbers that satisfied the equation.

It is these students who used the discovery this way whom I wish to discuss. Some of them spent several days working with various linear equations, using the patterns to find points, then checking by substituting into the equation. This was one of the earliest things the project ever did that seemed to fascinate children far beyond their normal degree of involvement with school.

My conjecture is that the important aspect of this was perhaps a combination of achievement in making their own discovery, competitive gratification vis-a-vis those classmates who did not make the discovery, autonomy of having set the task for themselves, some intrinsic esthetic or closure reward, and the existence of a verification that did not depend upon the teacher.

These were culturally deprived children, with a middle-class teacher and a middle-class curriculum. Previous observation had already convinced us that these children were at best *tolerant* of a schoolish learning that was *wrong* in every important respect. The children were always checking up, between school and social realities outside of school, and invariably found the school to be wanting.

The school taught you to speak the language incorrectly—for example, the school taught, 'It is I' in a world where the social reality was, 'It's me'—the school taught personal economics incorrectly, the school taught civics incorrectly, and so on. Nothing ever checked when you tried it out against actual reality. (For example, if a majority of people in the county don't like the sheriff, he won't be reelected—according to what they teach in school. But in reality, perhaps a majority of the people are not even allowed to vote—to use an example that is in our newspapers at the present moment.)

Here, in their own discovery, they had found something that *really worked!* If you tried it out, it checked out perfectly!

In any event, some of the children were captivated, and spent several days on the matter, asking to take graph paper home and so on.

(Incidentally, in the long run—i.e., over the year that we worked with this class—truancy decreased markedly, and parents reported their children taking an unprecedented interest in school—for example, by discussing it at mealtime.)

My second example is perhaps somewhat similar. We wanted to give children in grades 3-9 some experience with *variables* and with *the arithmetic of signed numbers.* It is one of our principles that such experience should always be provided *in a sensible mathematical context,* and (if at all possible) in a form which would permit a student to make one or more interesting discoveries. We believe that this procedure helps get the children in a frame of mind where they are always poking around looking for interesting patterns that may be lurking just beneath the surface.

I should emphasize that the discoveries in question will usually not be part of the basic purpose of the lesson, and it is not essential for a child to discover them. They are primarily a bonus for those who do discover them.

In the present instance, remember, we wanted the children to get some experience using *variables,* and working with the *arithmetic of signed numbers.* We consequently gave them quadratic equations to solve, beginning with

$$(\square \times \square) - (5 \times \square) + 6 = 0$$

and gradually progressing to harder problems, such as

$$(\square \times \square) - (20 \times \square) + 96 = 0$$

Now, at first the only method available to the student was, of course, trial and error. If he makes no discoveries, the student continues with this method, and gets full benefit from the basic part of the lesson; that is, he gets a great deal of experience using variables and signed numbers, and in a situation where he does not regard this as drill.

But—if the student discovers the so-called coefficient rules for quadratic equations, his use of trial-and-error can be guided to maximum efficiency. He has discovered a secret—and one which many of his classmates don't know. They may *never* know!

Torpedoing

In both of the previous examples, we make use of a technique which we call 'torpedoing.' After a student has discovered what he believes is the pattern for linear equations from working with

$$(\square \times 1) + 3 = \triangle$$
$$(\square \times 1) + 5 = \triangle$$
$$(\square \times 1) + 2 = \triangle$$

and so forth—*after* he is confident of his mastery, we unobtrusively slip in a problem like

$$(\square \times 2) + 3 = \triangle$$

He uses his pattern, he checks—and the numbers don't work! What shall he do?

With a little thought, he discovers there is a pattern here, also—indeed, there is a more general pattern of which he had discovered only a special case.

In a similar way, with the quadratic equations, we begin by using only unequal prime roots, so that one of the two coefficient rules (the product rule) is extremely obvious. Using it alone leads to easy solution of the equations, such as

$$(\square \times \square) - (\ 5 \times \square) + \ 6 = 0$$
$$(\square \times \square) - (12 \times \square) + 35 = 0$$
$$(\square \times \square) - (13 \times \square) + 22 = 0$$
$$(\square \times \square) - (\ 7 \times \square) + 10 = 0$$

and so on.

Here also, once the student is really pleased with his discovery, and with the new power it has given him, we confront him—unobtrusively and unexpectedly—with a variant problem which will tend to confound his theory.

In this instance, we slip in a problem having composite roots, instead of the prime roots the student had previously dealt with.

The product rule now seems to indicate more than two roots: for example, with

$$(\square \times \square) - (9 \times \square) + 20 = 0$$

many students will say the roots are

$$\{2, \quad 10, \quad 4, \quad 3,\}$$

Trial by substitution shows that this is wrong. Again, by persevering, the student finds that there is a broader theory, of which he had found only a narrower part.

Why Do We Like Torpedoing?

It may seem that what I have described is simple sadism, or How to Be One Up on Your Students Without Really Teaching. We *feel* that it is not so (although, unfortunately, it *can* be in the hands of a teacher who really *is* a sadist.)

Why do we use this technique of torpedoing some of our students' best theories?

It is important to realize that, at the outset, we don't know. We use the technique because, intuitively, it feels right.

But, in the years while we have been using this technique, we have, of course, discussed it often, and even made some analytic attempts at constructing an abstract rationalization for it.

Perhaps we like the technique because:

i) It gives the brightest students something to work on while the others catch up on more basic work;

ii) It is a friendly challenge from teacher to student, and students rise to such bait better than fresh-water fish do to flies;

iii) Perhaps Piaget's processes of 'assimilation' and 'accommodation' need to be practiced, and this is where you practice them;

iv) Or, to put point iii in less technical language, perhaps the security of a friendly classroom is the best place to gain experience in fixing up theories that used to work, but somehow don't seem to work any more—the classroom is a better place to learn this than, say, the political meetings of anti-integrationists or the radical right;

v) Then, too, it is worth learning that science does not deal in absolute truth. Sufficient unto each day are the theories thereof—and an irreconcilable contradiction may be discovered tomorrow! (Or, for that matter, a better theory!) If your theories work, make the most of them—but keep a wary eye out, just the same.

vi) Finally, this is one of the ways that we go about bringing history into the classroom. If one wishes to *understand* history, one must have some background of relevant experience. Since the history of mathematics is an unending story of trials, failures, break-throughs, temporary successes, new points of view, and so on, it is unintelligible to the person who has no background experience in trials, failures, break-throughs, temporary successes, revised points of view, and so on. Torpedoing theories in the classroom provides background experience that parallels important historical phenomena. Can you *realize* what the discovery of irrationals meant to the Greeks if you, yourself, start out with the sophisticated viewpoint of the 20th century?

The 'Crisis' Dilemma: Seeking the Unit Matrix

This next example involves a different kind of discovery situation. In the previous examples, the discoveries were merely optional bonuses added to the meat of the lesson; moreover, students discovering any secret patterns kept the secret to themselves, revealing their knowledge only indirectly, by using the secrets to solve difficult problems easily and quickly.

In the present example, all eyes are focussed upon a central problem which we wish to solve, if possible—or to recognize as unsolvable, if a logical argument shows that no solution can exist. If anyone finds an answer, he will announce his discovery at once.

Specifically, we look at the system of 2-by-2 matrices and ask how this new mathematical system compares with the familiar old system of rational numbers. One question is this: is there a 2-by-2 matrix which plays a role analogous to that of the rational number zero? The answer turns out to be that there is, and it is the matrix

$$\begin{pmatrix} 0 & 0 \\ 0 & 0 \end{pmatrix}$$

So far so good. Now—is there a matrix that plays a role analogous to the rational number *one*? Students invariably—and wisely, on the available evidence—guess

$$\begin{pmatrix} 1 & 1 \\ 1 & 1 \end{pmatrix}$$

but a quick computation shows that this is *not* satisfactory.

Here we have the dilemma: *IS* there any matrix that behaves like the integer one? Has our failure to find one been a symptom of the impossibility of the task, or have we merely failed due to personal reasons, not reasons of fundamental impossibility. Is it worth-while trying any longer? If so, how shall we proceed?

This has proved to be a consistently exciting lesson for fifth-graders, or anyone older who doesn't already know the answer.

Going Beyond the Data or the Task

In the filmed lesson entitled, "Graphing an Ellipse," at the end of a lesson on graphing

$$x^2 + ky^2 = 25, \quad \text{O} \leqq k,$$

a seventh-grade student (Debbie H., according to her nametag) asks: "Why couldn't you use matrices, and make a graph for k less than zero?"

Since these students have previously used matrices to introduce complex numbers, this is an ingenious and appropriate suggestion. The teacher does not immediately respond, but another student (whose nametag reads Lex) answers: "No, you can't, because you won't be able to graph matrices." This answer is essentially correct—but both of these remarks go beyond anything the teacher had planned or anticipated.

The teacher's contribution to this is mainly a genuine appreciation of the students' contributions—but we believe this is important. Children somehow act far cleverer when their cleverness is welcome and appreciated.

Do It Your Own Way: Kye's Arithmetic

Somewhat similar is this example. A third-grade teacher was introducing subtraction, with borrowing and carrying:

$$\begin{array}{r} 64 \\ -28 \\ \hline \end{array}$$

She said: "You can't subtract 8 from 4, so you take 10 from the 60 . . ."
A third-grade boy named Kye interrupted: "Oh, yes you can!

$4 - 8 = -4$

$$\begin{array}{r} 64 \\ -28 \\ \hline -\ 4 \end{array}$$

and $60 - 20 = 40$

$$\begin{array}{r} 64 \\ -28 \\ \hline -\ 4 \\ 40 \end{array}$$

and $40 - 4 = 36$

$$\begin{array}{r} 64 \\ -28 \\ \hline -\ 4 \\ 40 \\ \hline 36 \end{array}\ \text{."}$$

The teacher did nothing here to *solicit* originality, but when she was confronted with it, she *listened* to the student, tried to understand, and *welcomed* and appreciated his contribution.

This was an unusual, but actual, occurrence. The more common, 'traditional' response would have been to say: "No, Kye, that's not the way you do it. Now watch carefully and I'll show you . . ."

Where does this traditional rejection of his contribution leave Kye? He is given the feeling that mathematics is a stupid subject that never works out the way you'd expect it would. . . . Given enough experiences of this sort—and in a traditional situation he will be—Kye will probably transfer his interest and his energy to some other field of endeavor.

This example has always seemed to me to suggest the essence of good 'modern' teaching in mathematics, as opposed to 'traditional' teaching. In a phrase: *listen* to the student, and be prepared for him to suggest a better answer than any you know. The 'modern' teacher *actually learns from his students!*

Kye's algorithm for subtracting was an original contribution of a third grade boy. It is in many ways the nicest algorithm for subtracting that I have ever seen—and it was invented by a boy in the third grade.

The traditional teacher assumes from the outset that such a thing is impossible. Is it any wonder that the traditional teacher somehow never encounters such clever behavior from students? Like the spirits that move Oui-ja boards, such clever student behavior rarely appears before the eyes of those who don't believe in its existence.

Autonomy and Proliferation

In introducing graphical integration and differentiation to an eighth-grade class, we proceeded as follows: we obtained a print of the PSSC film "Straight Line Kinematics" which is an *expositional* treatment of these topics in relation to velocity and acceleration. The students could view this film whenever they wished—and as often as they wished. They were then given a shoebox full of simple equipment—the PSSC ticker tape equipment—to take home if they wished, and were asked to devise their own experiments and work up their own data. They later performed their experiments in school, at a session which has been recorded on film.

The effort and ingenuity that some students put into this went far beyond their normal effort for 'schoolwork.'

What Questions

In one instance, we give physical apparatus to students and ask them what *questions* might be worth studying about this apparatus. The point here is for the student to identify appropriate *questions*. An eminent mathematician, Professor McShane of the University of Virginia, has said that his favorite mathematics problem is stated in a textbook as follows:

A pile of coal catches on fire.

These seven words are the *entire* statement of the problem—in a *mathematics* book!

Now—what *mathematics* questions does this pose?

Leaving Things Open-Ended

Many mathematics problems occur at a stage in the child's life when he is not yet prepared to answer them. An honest use of logic seems to compel us to leave these questions open for the time being. The alternative would be to 'answer' them on the basis of authority—but we believe this would tend to make the child think that mathematics is based upon the pronouncements of authorities. We prefer to leave the question open—after all, aren't most scientific questions open at the present time? Or, perhaps, *always* open?

Example: Is

$$\square \times O = O$$

an *axiom* or a *theorem?* Ultimately the child will learn that it is a theorem:

$$\square + O = \square$$
$$O + O = O$$
$$\square \times (O + O) = \square \times O$$
$$(\square \times O) + (\square \times O) = \square \times O$$
$$\therefore \square \times O = O$$

This proof, however, involves some subtle and awkward points, so that the child cannot settle the question when it first occurs to him, and we leave the question open. The child knows that if he is ever able to *prove* the result, he will be able to classify it as a theorem. In the meantime . . . Who knows which it is?

Where Does It Come From?

The Madison Project approach to *logic* is, so far as we know, completely unique and unprecedented. Virtually every existing book on logic *tells* you what *modus ponens* is, *tells* you what the truth-table is for 'and,' 'or,' 'if . . . then,' and so on.

But why? Where does all of this come from, anyhow?

Given the transitory nature of scientific knowledge, we can hardly settle for facts which are static pieces torn out of the fabric of time past, time present, and time future.

Where does all of this logic come from? How would we go about making up our own logic if we wished to do so?

In order to answer these questions—at least, according to our own view of the answer—the Madison Project proceeds like this:

First, we ask children (grades 7, 8, 9, or older) to analyze statements of their friends, and to classify them as true or false. They realize that this is a vast oversimplification, for most ordinary statements in the ordinary world are neither true nor false; there's some truth in them, but one still has some possible doubts or reservations.

We then ask the children to focus on actual usages of 'and' and 'or,' and to record *as many different uses of 'and' and 'or' as they can find* by means of truth tables.

Ordinary language has a great many different uses of both words. One of my favorites—we might label it 'and$_1$,' in order to distinguish it from *other* uses, which can be labelled 'and$_2$,' 'and$_3$' and so forth—is this one:

Keep driving like that and you'll kill somebody.

A somewhat similar use of 'and' occurred prior to the 1964 election, in the radio admonition: Vote, and the choice is yours, Don't vote, and the choice is theirs. If you fail to register, you have no choice. Consider, also, the poem: "Laugh, and the world laughs with you, Weep, and you weep alone." The result of this activity is an extensive truth table with many different columns, headed 'and$_1$,' 'and$_2$,' 'and$_3$,' etc., and 'or$_1$,' 'or$_2$,' 'or$_3$,' and so on. Usually some students insist on moving into a more-than-two-valued logic, in order better to reflect nuances of meaning which seem to them to be present. Once we have collected together this large truth table, we have completed stage one, which might be labelled Observing the Behavior of the Natives.

We move next to the Legislative stage: In order to gain clarity, we agree on one single meaning for the word 'and,' we pick one column in the truth table to define this meaning, and we legislate that 'and' shall henceforth be used in that single sense only.

The study of logic proceeds further: after the Sociological stage, and the Legislative stage, we move on to the Abstract stage, and so forth—but the point is that these children have made up their own systems of logic. As a result, they know where logic comes from. In the same way, you understand Beethoven differently after you, yourself, have written some music of your own composition.

An Active Role, and Focussing Attention. Discovery of another sort is perhaps involved in teaching students (second-graders, say) to plot points on Cartesian coordinates. The teacher plots a few points, but the students learn more by imitation than by following a careful exposition. Learning by imitation of course involves a kind of discovery, since you must figure out how the teacher is doing it.

As David Page has pointed out, the best mathematics students have always learned by discovery—even when listening to a lecture, they are *actively thinking:* asking Why? Why not? How about doing it this way? Now what do you suppose he meant by that? Why can't you do it *this* way? and so on.

Games Using Clues. One of our most successful lessons goes like this:

Three students, working together, make up a rule. For example, Whatever number we tell them, they'll double it and add twenty. They do *not* tell us their rule; instead, we tell them numbers. They apply their rule to each number we tell them, and tell us each answer. It's our job to guess their rule.

There is a great deal of mathematics involved in this lesson—for example, the distinction between formula and function, and such properties of functions as linearity, oddness, evenness, rate of growth, and so on. But perhaps the main *discovery* aspect resembles closely Suchman's (1964) work on inquiry: By choosing wisely the numbers we tell them, we can get *clues* as to the nature of the function they are using. No single clue will usually be decisive, but a suitable combination will be.

A somewhat similar format has been used by David Page in his lessons on Hidden Numbers. A few numbers are written on a piece of paper, and the students are to guess the numbers from a set of clues. To make matters harder, *some of the clues are false,* so that the students must recognize *and make use of the contradictions* which are contained in the set of clues (grades 4, 5, 6, 7, 8, or 9).

Two Theoretical Interpretations

An extremely valuable approach to analyzing communications in class has been developed by Professor J. Richard Suchman. The data inside a student's mind at an instant in time can be classified (as an oversimplification, of course) into three categories: facts, unifying mental constructs (roughly, 'theories'), and applications. The possible communications can be diagrammed as follows:

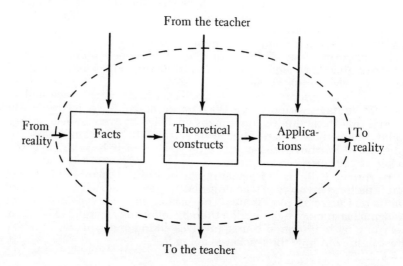

Using this analytical approach in interpreting classroom lessons can be very exciting and gratifying—which, as J. Robert Oppenheimer has remarked, is perhaps the most valid test of a theoretical approach. Because of limitations of time and space, I shall not attempt to illustrate the use of Suchman's diagrams, beyond remarking that on a Suchman diagram, discovery communications seem to appear as conspicuously horizontal channels, whereas expositional and rote communications appear as conspicuously vertical channels. Particularly in connection with the work of Piaget, Tolman, Lewin, Kohler, White, and Bruner, Suchman diagrams are a powerful analytical tool for the practitioner seeking a rationale for understanding what goes on in the classroom.

Another aspect of discovery experiences is being emphasized in the motivation studies by Richard de Charms (in press) in relation to the analytical dichotomy of the learner's perception of his role, which de Charms polarizes as 'origin' versus 'pawn.'

The Goals of Education

It is worth remembering that the life of a human being, or the interrelated lives of many human beings, are in reality unified wholes, not separated into pieces in any way whatsoever. When, in order to analyze life abstractly, we break off pieces by invoking *categories,* we do a violence to the reality whole that can impede comprehension as easily as it may, hopefully, facilitate it.

In particular, motivation is not separate *in reality* from perception, personality, learning, social interactions, or communication. Still more specifically, *reality* does not begin with a statement of *explicit goals.* If a teacher begins with an attempted statement of explicit goals, he does so by choice, and may easily do so in error, for his subsequent behavior may not reveal the same goals as his initial words did.

In our own Madison Project teaching, our original conscious goals were general and nonspecific: we wanted to find some of the best experiences with mathematics that children could have. To choose among alternatives we relied upon our intuitive assessments.

From this highly general purpose we spun out a sequence of specific activities in the classroom. In order to *discuss* these experiences, *after they were created,* we have attempted to work out some suitable analytical categories. In particular, we have identified what appear to be some of our probable goals. Here is a tentative list, surely somewhat incomplete:

i. We want to give students experience in *discovering patterns in abstract situations;*

ii. We want students to have experience in *recognizing potentially open-ended situations,* and in *extending open-ended situations by original creative work;*

iii. We want the students to be familiar with the basic concepts of mathematics, such as *variable, open sentence, truth set, function, Cartesian coordinates, mapping, isomorphism, linearity, matrices, implication, contradiction, axiom,* etc.

iv. We want the students to build up, in their own minds, suitable *mental* imagery (in the sense of Lewin, Tolman, Piaget, Leibnitz, Polya, et al) to permit them to perform mental manipulations involving the basic ideas of mathematics (such as *function, linearity, isomorphism, mapping,* etc., as mentioned above);

v. We want the students to acquire a modest mastery of the *basic techniques* of mathematics;

vi. We want the students to know the basic facts of mathematics, such as 7 + 3 = 10, -1 × -1 = +1, and so on;

vii. We want the students to possess considerable facility in *relating the various parts of mathematics one to another*—for example, using algebra as a tool in studying geometry, or recognizing the structure of the algebra of linear transformations of the plane into itself, and so on;

viii. We want the students to possess an easy skill in *relating mathematics to the applications of mathematics* in physics and elsewhere;

ix. We want the students to *have a real feeling for the history of mathematics,* derived partly from having been eye-witness observers (or participants) on the occasion of mathematical discoveries.

We regard the preceding nine points as intellectual matters; but they must be accompanied by some emotional or value goals, namely:

x. We want the student to know that mathematics *really and truly is discoverable* (something few people believe);

xi. We want each student, as part of the task of knowing himself, to get *a realistic assessment of his own personal ability* in discovering mathematics;

xii. We want the students to come *to value 'educated intuition'* in its proper place;

xiii. We want the students *to value abstract rational analysis* in its proper place;

xiv. We want the students—as much as possible—*to know when to persevere, and when to be flexible;*

xv. We want the students to have a feeling that *mathematics is fun or exciting, or worthwhile.*

The preceding goals do not sound like the goals of a traditional arithmetic or algebra program. They are not.

Space and time do not permit us to pursue the matter, but implicit beneath all Madison Project work is the notion that *education* and *training* are different, that education is for people, and training is for electronic machinery (which usually doesn't need it anyhow)—indeed, all repetitious routine tasks are basically nonhuman.

One example of the distinction—a tragic and highly suggestive example—will have to serve where many might be cited:

A few years ago in a hospital nursery in Binghamton, N. Y., the formula for new-born babies was made with salt instead of sugar. By the time the error was discovered, a dozen or so babies had either died or suffered severe and irreparable brain damage.

Now,—who or what was at fault? Many babies—indeed, virtually all of them—*simultaneously* developed feeding problems. Some of the *mothers*—but none of the nurses!—tasted the formula and complained that it was unusually salty.

No nurse heeded either clue. The nurses had been trained to soothe a new mother's anxieties, lest they be passed on to the infant and create feeding problems.

Those of us who have children of our own know how much this episode meant to the parents of the babies involved.

New York State has responded by passing a new law, which, I believe, makes it illegal for a hospital to store salt in a room in which nursery formulas are prepared.

This is a significant step in the wrong direction.

Every time we attempt to by-pass human resourcefulness—by laws, rote training, or otherwise—we move toward, and not away from, the unintelligent behavior of the nurses who were trained but not educated. The response of the *mothers* was more appropriate than that of the nurses, but the nonadaptive blind weight of authority decided the outcome in favor of the 'trained' nurses.

I think it important for every teacher always to remember that *he, the teacher, does not know the right answer or the right response*—he can only hope that, when the time comes, his former students will respond appropriately.

Those theorists who study education are on shaky ground indeed if their analysis assumes that they can separate right answers from wrong answers. If the matter in question is trivial, perhaps they can; but if it is important, they surely cannot.

The present emphasis on creativity and divergent thinking would never have occurred—and should never have occurred—but for the fact that we had gone all too far down the road labelled training, and had, surprisingly, lost sight of education.

Exercises: Discovery in the Teaching of Mathematics

1. Consider Davis' example of plotting $y = 3x + 2$. Identify the strengths and weaknesses you see in that lesson. Give another example. Would you use it in your class? Explain.

2. What is torpedoing? Make up your own example. What psychological framework can you provide from the other readings to support the procedure? Explain. How does it affect children? Beneficially? Adversely? Why?

3. Many beginning teachers believe that using a discovery lesson is easy because it requires no preparation. They think that the teacher can simply just go where the children's questions and work lead. Explain why this is fallacious reasoning. Use an example not given by Davis.

4. What can a teacher do to avoid the "No, Kye, . . ." response in his own teaching? Why is it so hard to avoid?

5. Many teachers reject discovery teaching because it takes extra time when there is so much material that must be covered in any one particular year. Is there an answer to this criticism? Explain. How does the problem relate to objectives?

6. Apply the Suchman diagram to a mathematical topic. Fill in each of the boxes and describe the procedures to be used to achieve each of the arrows.

7. Make a list of both short and long term goals for a topic in mathematics. Choose those things which you really believe are most important. Can you identify some as being especially appropriate for discovery and others that are especially appropriate for exposition? Explain.

Learning
by Discovery*

Gertrude Hendrix

The issue of learning by discovery and of teaching for learning by discovery is beclouded at present by the fact that each of *three* very different procedures is being called *'The* Discovery Method.' They are the inductive method, the nonverbal awareness method, and the incidental method.

The Inductive Method

Many people think that 'the discovery method' is just another name for the inductive method. The inductive method is nothing new in mathematics education. Colburn's book on teaching arithmetic by this kind of approach[1] was first published in the early 1820's. The companion volume on algebra[2] came a few years later. This method was strongly advocated in reports of the National Committee on Reorganization of Secondary School Mathematics in 1918 and 1923. That committee on the Mathematical Association of America drew heavily on earlier recommendations of E. H. Moore in the United States, Perry in England, and Felix Klein in Germany—mathematicians who were writing of these things around the turn of the century. Again in the early 1940's, the Joint Committee of the National Council of Teachers of Mathematics and the Mathematical Association of America, in a report published as a yearbook of the National Council,[3] strongly advocated the inductive method. These recommendations, however, were acted upon by such a small minority of teachers of mathematics that most people seeing the inductive method in action today think they are seeing something new. Furthermore, in clarity and productive outcome it is so far superior to an authoritarian "tell'm and drill" approach that, to someone whose school experience has been dominated by a formal approach, the inductive

Gertrude Hendrix, "Learning by Discovery," *The Mathematics Teacher,* May 1961, pp. 290-299.

*"Learning by Discovery" is likely to be misinterpreted by readers who have neither witnessed nor experienced many examples of the unverbalized awareness stage in abstract learning. The paper was written to be a part of the manual for "Teaching High School Mathematics, First Course," a series of fifty teacher training films distributed by Modern Learning Aids, 1212 Avenue of the Americas, New York, New York, 10036. These films contain many demonstrations of the instructional approach presented in "Learning by Discovery."

[1]Warren Colburn, *First Lessons; Intellectual Arithmetic upon the Inductive Method of Instruction,* new edition, with an introduction to written arithmetic by his son, Warren Colburn, and an introduction by George B. Emerson (Boston: Hilliard, Gray, Little and Wilkins, 1863). [Original edition about forty years earlier; others in 1849, 1858.]

[2]Warren Colburn, *An Introduction to Algebra upon the Inductive Method of Instruction* (Boston: Cummings, Hilliard, and Company, 1826).

[3]National Council of Teachers of Mathematics, *Fifteenth Yearbook,* (New York: Bureau of Publications, Teachers College, Columbia University, 1940). Cf. item (c) p. 57.

method seems the final answer to all pedagogical difficulties. Hence, someone just now beginning to think seriously about theory of instruction is likely to look upon the inductive method as the ultimate answer. *The fallacy in the inductive method lies in its confusion of verbalization of discovery with the advent of the discovery itself.*

The separation of discovery phenomena from the process of composing sentences which express those discoveries is the big new breakthrough in pedagogical theory. In some cases, correct verbalization of a discovery *does* emerge immediately after the dawning of awareness, but these cases are rare. They are confined to learners with unusual powers of correctness and precision in the use of a mother tongue, and they are confined to situations in which the learner happens to possess all the vocabulary and rules of sentence structure needed to formulate the new discovery. In most cases, for evidence that an inductive discovery has occurred, a teacher must be able to recognize responses or behavior other than a linguistic formulation of the discovery. The widespread confusion of sentence formation with the discovery process itself has been responsible for much of the frustration, discouragement, and ultimate abandonment of the inductive method on the part of many teachers who have attempted to master it. As one head of a teachers' college department of education has put it, "But the only ones who can state a generalization from examining examples are the same exceptionally bright ones who could have learned to apply the generalization from studying someone else's statement of it in the first place!" As a matter of fact, the only teachers who have been really successful with the inductive method are the rare ones who have acquired the knack of making the learner *aware* of generalizations before calling for verbalizations. They have been doing "right things for wrong reasons," and the subsequent untimely verbalizations have only partially marred the outcomes.

If the writer may be permitted so personal a statement, I should like to record here that it was while in pursuit of greater skill with the inductive method that I first became conscious of the nonverbal awareness stage in discovery learning. From concentrated work in helping student teachers to acquire skill in this method, I had become very conscious of three bad mistakes which beginners are likely to make when they plan and attempt inductive teaching:

1. They begin to call for generalizations, exerting anxious pressure on their classes, long before the students have noticed any basic similarity among the examples the teacher has presented. The discussion degenerates into a guessing contest during which the students try to find out "what it is the teacher wants me to say."
2. The teacher often calls for statements of generalizations when the students do not possess the vocabulary and rules of sentence formation necessary for a precise verbalization of the generalization, even if they have "seen it."
3. The teacher often confuses generalizations, which had to be discovered in the first place and, hence, are appropriate subject matter for rediscovery, with situations in which the generalization is arbitrary—something that is merely a matter of definition.

A frequent costly consequence of item 2 above is a teacher's yielding to a temptation to accept a faulty answer, an answer which is either false or nonsense. Spontaneous verbal responses at moments of discovery are usually expressive of emotion only. Often they are ejaculations, referentially haphazard. Sometimes they are in sentence form but seem to be utter nonsense if repeated and examined. Or, still worse, they may be plausible, *interpretable* sentences which state false generalizations. Spontaneous verbal fragments often provide clues from which a clever teacher can infer what is in a student's mind, and for that purpose they are valuable. But, unless the teacher is a tutor working with only one pupil, encouraging such responses can be very damaging to the learning under way in *other* members of the class. Even in the tutoring situation, an incorrect spontaneous verbalization of a discovery—a sentence which really says something else—must be ignored and forgotten as soon as possible. Accepting it as a "nice try" can cause an involuntary interpretation of the false sentence to prevail over the correct discovery. If *any* recognition is accorded to such a response, the teacher must stick with it until counter examples have revealed to the whole class that the sentence must be discarded. By this time the class has shifted to an exercise in language composition, which cannot be abandoned tactfully. Pushing on through to a correct verbalization at such a time usually demands a long, laborious digression. By the time the point is clear, most of the zest for the original learning has evaporated; and the production of the correct sentence, accepted at long last, may even destroy the anticipatory attitude with which ensuing applications of the generalization should be approached. (Experimentation has established that this effect of finality and detachment is avoided if verbalization of a generalization is postponed to a later lesson. At that time the linguistic formulation of things already "known" can be undertaken as an end in itself.)

The big problem in the inductive method has been to find some way of telling when students are far enough along in an inductive approach for it to be *fair* to ask them to state a generalization. It was while a search was under way for a criterion by which to identify this stage in each inductive lesson that an event came along which led to my identification of the nonverbal awareness stage in discoveries of concepts and generalizations. This happened in 1937.

As early as 1946, clinical experts in nondirective psychological counseling were pointing out that frequently there were very significant changes in behavior based on unverbalized insights, and that the effectiveness of an interview could be destroyed by premature pressure for verbalization. The counseling experts were *not* telling us, however, how to detect when it was *not too soon* to press for verbalizations. When I designed my first experiment[4] for research on this question, for one group of subjects the teaching was terminated at the nonverbal awareness stage simply to call attention to the fact that that stage existed, that there *was* a way to tell when a learner was aware of a generalization which he might be asked to state. At the time that I designed this part of the experiment, it did not occur to me that as far as transfer power was concerned, the *whole* thing was there as soon as the nonverbal awareness had dawned. No one was more shocked than I was to find that not only did the learners who completed correct verbalization of the discovery do

---

[4]Gertrude Hendrix, "A New Clue to Transfer of Training," *Elementary School Journal*, XLVIII (December 1947), 197-208.

no better on transfer tests than those for whom the teaching was terminated at the nonverbal awareness stage, but also that the verbalization of the discovery seemed actually to have diminished the power of some persons to apply the generalization. Charles Hubbard Judd in his conscious generalization theory of transfer had considered 'generalizing' as synonymous with 'composing a sentence which states the generalization involved'.[5] But that simply is not the case; unless Judd's definition of 'generalizing' is changed, we can no longer say that generalizing is the primary generator of transfer power.

Since 1946, the problems have become:
a) How can teaching be planned and executed so that necessary verbalizations of discoveries are accomplished *without* damage to the dynamic quality of the learning itself?
b) What is there about the nature of language and the process of acquiring a language, and what goes on when one uses language that could explain (and, hence, help us to avoid) detrimental effects of verbalization?

Very soon a tremendously important by-product of this search began to appear: The discovery process itself is so exhilarating (to both children and adults) that it becomes its own motive in academic work. As a solution to the motivation problem for children in a free society whose existence is threatened by rapidly increasing scientific and technical power in the land of an enemy, the promotion of successful learning by discovery has no parallel. But such teaching is an art with a difficult technique. Furthermore, it is dependent upon sound and clearly written textbook materials. Ambiguities, inconsistencies, "fuzzy" writing tolerated for centuries—all these must be eliminated from science and mathematics courses. (Another field in which such stumbling blocks abound is that of grammatical construction of word languages. The unifying concepts common to all languages are now revealed by developments of the last half century in symbolic logic, but as yet these developments have not been utilized in a reorganization of grammar.)

The Nonverbal Awareness Method

This method has been described incidentally in the above discussion of the inductive method. It will be further explained, perhaps, in the illustration developed in the second part of this article. This is the method of teaching which has been found most successful with the UICSM Mathematics Project materials. The materials are written to promote use of the method with a minimum of difficulty for a teacher to whom both materials and method are new.

The Incidental Method

This is an approach widely promoted during the Progressive Education era.

It is the method in which the school sets the stage for many

---

[5]"The human power of generalization is so intimately related to the evolution of language that the two cannot be thought of as existing separately. . .language is the chief instrument of generalization. . . ." C. H. Judd, *The Psychology of Secondary Education,* (Boston: Ginn and Co., 1927).

experiences, which are usually being built around some central problem or project. Those who advocated this method knew that only generalizations which emerged from experiences were likely to play a dynamic part in the learner's later behavior and problem-solving activity. But all too often they took no responsibility for seeing that instances of the same generalization came along close enough together for the learner to become aware of either concepts or principles. With no underlying explanatory theory, those educators simply did not realize that most learners under such conditions seldom "broke through" to awareness of abstractions. Such a school program was doomed to triviality, except in the hands of a teacher who did many additional things, even though he did them for unsound reasons. Many persons today associate this triviality of results, from many of the so-called 'activity programs' with learning by discovery. They thus reject anything called 'discovery method'.

Illustrations

The three separate and very different teaching procedures, each referred to by its proponents as '*the* discovery method,' can be illustrated by reference to ways of teaching a multiplication rule for quotients. All three of the so-called discovery techniques are in contrast to any approach through initial statement of a rule. Just to get that approach out of our minds let's examine two possible statements of such a rule, one verbal, the other mathematical:

1. For each first number, for each second number not zero, for each third number, for each fourth number not zero, the quotient of the first number by the second number multiplied by the quotient of the third number by the fourth number is the product of the first number by the third number divided by the product of the second number by by the fourth number.

2.

$$\forall_a \forall_{b \neq 0} \forall_c \forall_{d \neq 0} \, \frac{a}{b} \times \frac{c}{d} = \frac{ac}{bd} \, .$$

Such sentences are avoided in traditional textbook approaches by resorting to garbled mixtures of arithmetic and metamathematics—rules which abandon references to numbers and tell one how to manipulate figures to get an answer. The very existence of such language as '. . . the product of the numerators over the product of the denominators' traces to confusion between numerals and the entities which numerals represent. Mathematicans have become aware of entities and relations between the entities; they have then named, or symbolized, the entities and described the relations between them; and then they have thought that they were giving us the essence of their discoveries when they gave us the symbols and sentences. In the resulting desperation to make things clearer, they have composed verbal descriptions of *their symbols and what they do with them,* a sorry substitute for making those with whom they are trying to communicate aware of the concepts and generalizations they themselves have in mind on a nonverbal level. Stating an arithmetic generalization concerning products of quotients requires sentences about numbers. If a mathematician says what he means, he needs sentences of the types illustrated in (1) and (2). In contrast, let us see how discovery lessons based on each of the three interpretations of 'discovery method' outlined above would proceed.

Prerequisite to all three of the approaches are these previous learnings:

1. When a rectangle is divided into six congruent squares, each square is called 'one-sixth of the whole'; or, in general, for each counting number n, if a rectangle is divided into n congruent squares, then each square is called 'one-nth of the whole rectangle'.

2. When a rectangle two units wide and three units long is divided into unit squares, the resulting six squares can be thought of as two of three, or as three of two, that is, two rows of three squares each, or three columns of two squares each. That is, for each counting number m, for each counting number n, a rectangle m units long and n units wide is made up of m of n squares or n of m squares.

3. The number of unit squares in any rectangle of the kind described above is mn, that is, multiplication is the arithmetic process by which one finds the area-measure of a rectangle when he knows its length-measure and width-measure.

Inductive Method

The class reviews the three items above, probably in the reverse order from that in which they are stated above. [It would be a more natural quasi-research approach if a problem to be solved had been stated first, for example:

What is ½ of ¾?

Unfortunately, things are not usually done in this order. Hence, the

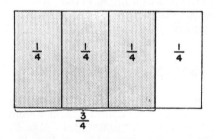

Figure 1

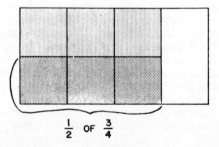

$$\frac{1}{2} \quad \text{OF} \quad \frac{3}{4}$$

Figure 2

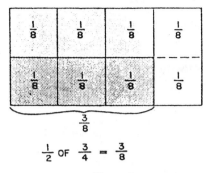

Figure 3

prerequisite review is imposed authoritatively by the teacher, instead of being something thought of as a way of shedding light on a new, unsolved problem. Conducting a review before stating the problem injects a false and unnatural element into the discovery process, and robs the children of experience in an important technique of research.]

Second, if not earlier, the problem is stated, and the teacher helps the student to find an answer to the question, "What is ½ of ¾?" by making appropriate drawings (Figures 1, 2, 3) and then counting.

First, a rectangle is divided into fourths and three of them are shaded. Next, the shaded part is divided into halves, and cross shading is used to identify one-half of three-fourths of the original rectangle. Finally, each part of the cross-shaded section is found by counting to be one-eighth of the original rectangle. Then, by counting, the children see that one-half of three-fourths of the rectangle is three-eights of the rectangle. So, the teacher records on the chalkboard at one side, possibly under the heading 'Data':

$$1/2 \text{ of } 3/4 = 3/8 .$$

Referring back to 'two of three' and 'three of two' as synonyms for '3 × 2' and '2 × 3,' all agree that it should be all right[6] to restate the last problem as:

$$3/4 × 1/2 = 3/8 .$$

The period at the end of the first line of data is erased and the line is changed to this restatement of the problem:

$$1/2 \text{ of } 3/4 = 3/8 , \qquad \text{or } 3/4 × 1/2 = 3/8 .$$

This process of making and examining appropriate drawings is repeated for at least three more examples, but usually not more than that. By this time, the recorded data may look something like this:

$$1/2 \text{ of } 3/4 = 3/8 , \qquad \text{or } 3/4 × 1/2 = 3/8 .$$
$$2/3 \text{ of } 1/5 = 2/15 , \qquad \text{or } 1/5 × 2/3 = 2/15 .$$
$$3/4 \text{ of } 3/5 = 9/20 , \qquad \text{or } 3/5 × 3/4 = 9/20 .$$
$$3/7 \text{ of } 2/3 = 6/21 , \qquad \text{or } 2/3 × 3/7 = 6/21 .$$

[6]In the UICSM materials the "times" sign, '×,' means *multiplied by*. That is, in '¾ × ½,' the number named second is the multiplier.

A teacher alert to the structural properties of the set of numbers of arithmetic will also have, for example:

$$3/4 \text{ of } 1/2 = 3/8, \qquad \text{or } 1/2 \times 3/4 = 3/8.$$

The differences between the experimental procedure for this exercise and the drawings in Figures 1 to 3 will have been illustrated on the board as follows with Figures 4 to 6:

New Problem: ¾ of ½ = ?

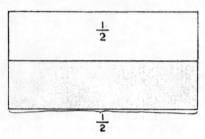

Figure 4

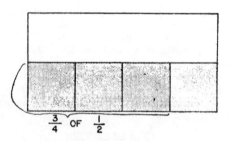

Figure 5

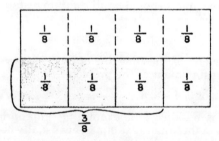

Figure 6

So ¾ of ½ = ⅜, or ½ × ¾ = ⅜, and a corresponding entry is included in the table of data.

The next part of the inductive lesson is usually introduced by the teacher with some remark, such as "Now, let's see how we could have found the answers just by looking at the fractions." The children are told to examine each pair of fractions and the fraction which names the product, and to try to tell how they could have found the product without drawing the picture. In the inductive method, *the ability to "tell how to do it" is the criterion by which the teacher recognizes that discovery has taken place.*

As a matter of fact, children at this stage in arithmetic do not have the linguistic machinery for stating precisely what they mean when they *have* discovered the rule. The inductive method forces them into saying such things as "Well, you multiply the top numbers, and you multiply the bottom numbers, and then you write the top answer over the bottom answer."

A teacher's insistence on more polished statements may call for something like "Well, to get the numerator of the answer, you multiply the numerator of the first fraction by the numerator of the second fraction; then. . . ."

If the teacher is one of those rare ones who is trying to keep clear the distinction between numbers and numerals, he must then go back and remind the student that a fraction is not a number; it is a *name* for a number. So the student must revise his answer to say that the numerator of the answer is a numeral for the product of the numbers named by the numerators of the two original fractions, etc. All this time, the teacher keeps insisting that "If you can't tell it, you don't know it."

Finally, a statement acceptable to the teacher is achieved. Each student may write the new rule in his notebook, or he may be asked to turn the page in his book, where he will find a rule already stated. (Of course, some members of the class may have looked ahead already and found out what they were supposed to be saying.) The students are then considered ready to work a set of practice exercises.

When well done, this method does give each student what is sometimes called 'an intuitive crutch,' a device upon which he can fall back; he can repeat the experiment whenever he finds himself in doubt about the rule. Security and independence are provided, and, in contrast to a tell-them-and-drill approach, the inductive lesson seems a revelation in pedagogy. A well-presented authoritarian approach is very satisfying to someone who *is already aware* of the ideas being presented; but for a learner to whom the ideas are new, the authoritarian method at its best is a *feeble* best in comparison with the inductive method. And *the inductive method at its best* is a feeble best in comparison with a genuine discovery approach which promotes and recognizes a nonverbal awareness stage in creative thought.

The Nonverbal Awareness Method

Let us contrast the procedure called the inductive method with a discovery lesson which follows more closely a pattern of genuine research. The goal, as in the preceding development, is the "multiplying fractions" rule.

In this approach, statement of a problem *must* precede the developmental review leading to the physical space drawing experiment by which an answer is found. From then on, the lesson may proceed exactly as in the inductive method up to a certain point. Then a highly significant, sharply defined difference appears. In *this* kind of discovery approach, as soon as each child has mastered the drawing-and-counting device for finding products of pairs of quotients, he is given a list of problems to work. As far as the student knows, he is to work every problem on the list by making an appropriate drawing. The class activity thus shifts to individual seat work. What is not done before the bell rings is to be finished for homework. The problems on the list should be arranged to promote interesting observations. The work sheet should provide a space for recording the answers to the problems. (At least two pairs of consecutive examples will illustrate the commutative principle for multiplication.) To give pause to the conscientious little task-doer who might work the whole list by making drawings just because that is what the directions (or the teacher) said to do, examples like the one below should be included:

$$17/21 \times 11/19 = ?$$

Students accustomed to this kind of discovery approach will, of course, be on the lookout for a short cut. Furthermore, from previous experience with the course and their teacher, pupils will know that finding and using a short cut is something that will be highly approved. Until this experience has been built up, the directions for the exercise must contain some challenge, such as "If at any time you think you know what an answer is going to be *before* you make the drawing, please write down the answer and raise your hand." When a hand goes up, the teacher should go to the student, congratulate him, or advise him to check his hunch by another experimental drawing. It must be pointed out, however, that this direction injects a false note into the stage setting for true research. It is simply a device by which a teacher helps a class to get its first experiences with this kind of learning.

In such a lesson, *how does the teacher know when a discovery has taken place?* Suppose he is walking about the class looking over shoulders. A sudden start, a flush of excitement, and a student begins writing answers almost as fast as he can put them down. When a student begins to write correct answers without making the drawings, we have evidence of the advent of awareness. Awareness of what? Awareness of the "multiplying fractions" rule.

If the child were not aware of it, he couldn't be using it. To call a halt in his progress at this point for a long, tedious verbalization attempt is damaging in several ways. In the first place, it belittles the exhilarating accomplishment of the actual discovery. Whether he can "tell it" or not, the *child* knows very well that he *has* something.

Then, in the second place, the verbalization attempt at this grade-school level is doomed to frustration. Numerical variables, universal quantifiers, and the zero restrictions on domains are all necessary to a precise formulation of the generalization the child has just discovered. The most that a teacher can do is to accept a garbled verbalization of what was at the beginning a clear, dynamic insight. The more sensitive a child is to precision in the use of his mother tongue, the more damaging this process can be to him and to the learning which a moment before was so clear on the nonverbal awareness level. The more skillful the child is in interpreting his mother tongue, the higher the probability that a literal interpretation of his own incorrect statement will impinge itself upon him. If this literal interpretation of his incorrect statement contradicts the awareness that was already there, the original learning is mutilated.[7]

---

[7] I use the word 'mutiliated' deliberately and almost angrily. My feeling about this issue has been built up from several years' experience in seeing this thing happen to children over and over again.—The Author

In a lesson of this kind, the child who notices that his neighbor is writing correct answers without making drawings is electrified into a search for "what's going on." Knowing that one of his classmates has discovered a short cut has a much more highly motivating effect upon him than a search for short cuts suggested by the teacher. The child who does not discover the rule before the end of the class period still has his chance during his homework period—if well-meaning parents do not step in and spoil it all for him!

On the following day the teacher presents other examples, such as:

$$5/3 \times 7/8$$
$$6/5 \times 9/4$$
$$6/2 \times 21/7$$

In other words, *over*generalization and incomplete generalization are prevented by the introduction of *examples* which correct possible false impressions, *not* by verbalizing the rule at this stage of the game and conducting a sophisticated search for false instances. (The more sophisticated procedure is available and appropriate a few years later when these same children will have acquired the necessary linguistic equipment.)

I do not wish to belittle the tremendous improvement in instruction which results from pursuing the inductive method, in contrast to authoritarian procedures. But those who think teaching for discovery *is* the inductive method are still blinded to this second tremendous breakthrough in theory of instruction. It is recognition of the nonverbal awareness stage in inductive learning that converts the classroom experience into that of actual discovery, the kind of thing that promotes a taste for and a delight in research.

The Incidental Method

In contrast to both of the instructional procedures outlined above, let us think about the teacher who trusts that the "multiplying fractions" rule will be discovered incidentally.

In one project a child will need to find the cost of a sheet of gold leaf to line a jewelry box, for instance, when he knows the cost per square foot and knows that the sheet must be one-half foot wide and three-fourths foot long. He is taken through the drawing experiment until he finds an answer. No other instances of multiplying one quotient by another come along in *that* project. Six weeks later (or six months, or a year) in connection with another project, he has another occasion to find the product of two quotients. Again the physical space drawings are used, the answer is found, and the project proceeds. All pressure from teacher and group is directed toward finishing the project. (Anyone who becomes intrigued by the multiplication process itself and digresses to investigate will be frowned upon by many of the educators advocating this kind of schoolwork.) From such examples widely distributed over time and space, the probability that the generalization will emerge is almost zero. Only a little genius who has somehow escaped with his taste for contemplation undestroyed will recall such widely scattered experiences all at the same time, and hence become aware of the mathematical generalization underlying them. Critics of education who confuse this kind of school program with everything else referred to as 'learning by discovery' are justified in opposing a discovery method. Much as I, the writer, abhor authoritarian presentations of sentences which express generalizations, the words *later* to be clarified by examples, I would prefer falling back upon that relatively deadly and deadening procedure, rather than to rely upon important basic principles "coming clear" from experiences incidental to so-called 'project activity'.

So much for the status of the discovery method in educational parlance and educational literature at the beginning of the decade to be known as 'the sixties'! It is high time, however, for another category to be added to the list when discovery in *mathematics* is under discussion. Even the instructional procedure described under the nonverbal awareness method is not properly called '*the* discovery method'. It is possible to arrive at discovery by pure deduction, that is, by manipulating sentences according to logical rules. It is not likely, but neither is it impossible, that the "multiplying fractions" rule would be discovered by derivation; most deductive proofs are checks for sentences first arrived at inductively; but *some* very important principles, both in mathematics and in physical science, *have* been discovered by deduction. A good example of something in secondary-school mathematics usually discovered (that is, *re*discovered) in this way is the rule for finding the measure of one side of a triangle in terms of measures of the other two sides and the cosine of the measure of their included angle. This is a case in which one begins with some sentences which express observations of a drawing. Starting with these sentences as premises, pure deduction takes over and out comes the desired formula. This last sentence, discovered deductively, is then interpreted in terms of how the variables were used in the original drawing. This interpretation of a final sentence in the deductive sequence reveals the desired geometric rule. This is a discovery process just as truly as is the inductive leap that takes place in the nonverbal awareness type of discovery. How do we know that both are discovery? Because a person can do them *all by himself.* The thing learned is not received by communication.

Much of the confusion permeating discussions of learning by discovery comes from failure to distinguish discovery from communication. One does not *discover* a concept or a principle in a book. He may discover there a *name* of a concept, a definition, a *sentence* which states a principle, etc., but before he "has" anything, he must engage himself in the process of interpreting the words and sentences of the author, that is, he must complete a communication process. Communication[8] usually plays an important part in setting the stage for discovery in the teaching of mathematics, but that part is quite different from actually communicating the new thing to be learned.*

---

[8]A rationale of communication adequate for explaining this distinction has emerged from the nonverbal awareness research at the University of Illinois; the author hopes to make the theory available in print soon.

*The writer acknowledges with gratitude many helpful suggestions from the following persons who read and criticized an earlier version of this article: Max Beberman, Alice G. Hart, Kenneth B. Henderson, R. Stewart Jones, David A. Page, and J. Richard Suchman, University of Illinois; Merrill B. Hill, Utah State Department of Public Instruction; and Stanley M. Jencks, University of Utah.

Exercises: Learning by Discovery

1. Choose a single lesson from mathematics not given in the article and use it to illustrate each of the following:
   (a) the inductive method
   (b) the nonverbal awareness method
   (c) the incidental method.

2. Which of these methods given by Hendrix do you believe best typifies Davis' approach to discovery? Explain.

3. Hendrix is critical of certain practices of teachers. What are they? Explain Hendrix's displeasure. What support can you find from the other readings for Hendrix's position? What psychological positions do not support Hendrix's position? Explain.

4. Discovery teaching has been criticized as a dishonest game of "guess what's on my mind." What can teachers do to avoid this difficulty?

5. Design a lesson following carefully the recommendations of Hendrix. What are the strengths of such a lesson? The weaknesses? Would you use it? Why or why not?

Some Psychological and Educational Limitations of Learning by Discovery

David P. Ausubel

Learning by discovery has its proper place among the repertoire of accepted techniques available to teachers. For certain purposes and under certain conditions it has a defensible rationale and undoubted advantages. Hence the issue is not whether it should or should not be used in the classroom, but rather for what purposes and under what conditions. As in the case of many other pedagogic devices, however, some of its proponents have tended to elevate it into a panacea. Thus, because many educators are tempted unwarrantedly to extrapolate the advantages of this technique to all age levels, to all levels of subject-matter sophistication, to all kinds of educational objectives, and to all kinds of learning tasks, it is important to consider its psychological and educational limitations. What doesn't learning by discovery do? What kinds of objectives can't we hope to accomplish by using it? When isn't its use appropriate or feasible? For what age levels or levels of sophistication isn't it suitable?

Problem-Solving is Not Necessarily Meaningful

The first psychological qualification I wish to propose disputes the widely accepted twin beliefs that, by definition, all problem-solving and laboratory experience is *inherently* and *necessarily* meaningful, and all expository verbal learning consists of rotely memorized glib verbalisms. Both assumptions, of course, are related to the long-standing doctrine that the only knowledge one *really* possesses and understands is knowledge that one discovers by oneself. A much more defensible proposition, I think, is that *both* expository *and* problem-solving techniques can be either rote or meaningful depending on the conditions under which learning occurs. In both instances meaningful learning takes place if the learning task can be related in nonarbitrary, substantive fashion to what the learner already knows, and if the learner adopts a corresponding learning set to do so.

It is true that by these criteria much potentially meaningful knowledge taught by verbal exposition results in rotely learned verbalisms. However, this rote outcome is not inherent in the expository method per se, but rather in such abuses of this method as fail to satisfy the criteria of meaningfulness. Some of the more commonly practiced and flagrantly inept of these abuses include "premature use of verbal techniques with

David P. Ausubel, "Some Psychological and Educational Limitations of Learning by Discovery," *The Arithmetic Teacher*, Vol. 15 (May, 1964): 291-302. ©1964 by the National Council of Teachers of Mathematics. Used by permission.

cognitively immature pupils; arbitrary presentation of unrelated facts without any organizing or explanatory principles; failure to integrate new learning tasks with previously presented materials; and the use of evaluation procedures that merely measure ability to recognize discrete facts and to reproduce ideas in the same words or in the identical context as originally encountered." [1, pp. 23-24.]†

Actually, a moment's reflection should convince anyone that most of what he *really* knows and meaningfully understands, consists of insights discovered by *others* which have been communicated to him in meaningful fashion.

Quite apart from its lack of face validity, the proposition that every man must discover for himself every bit of knowledge that he really wishes to possess is, in essence, a repudiation of the very concept of culture. For perhaps the most unique attribute of human culture, which distinguishes it from every other kind of social organization in the animal kingdom, is precisely the fact that the accumulated discoveries of millennia can be transmitted to each succeeding generation in the course of childhood and youth, and need not be discovered anew by each generation. This miracle of culture is made possible only because it is so much less time-consuming to communicate and explain an idea meaningfully to others than to require them to re-discover it by themselves.

There is much greater reluctance, on the other hand, to acknowledge that the aforementioned preconditions for meaningfulness also apply to problem-solving and laboratory methods. It should seem rather self-evident that performing laboratory experiments in cookbook fashion, without understanding the underlying substantive and methodological principles involved, confers precious little meaningful understanding, and that many students studying mathematics and science find it relatively simple to discover correct answers to problems without really understanding what they are doing. They accomplish the latter feat merely by rotely memorizing "type problems" and procedures for manipulating symbols. Nevertheless it is still not generally appreciated that laboratory work and problem-solving are not genuinely meaningful experiences unless they are built on a foundation of clearly understood concepts and principles, and unless the constituent operations are themselves meaningful.

Two related strands of the Progressive Education movement—emphasis on the child's direct experience and spontaneous interests, and insistence on autonomously achieved insight free of all directive manipulation of the learning environment—set the stage for the subsequent deification of problem-solving, laboratory work, and naive emulation of the scientific method. Many mathematics and science teachers were rendered self-conscious about systematically presenting and explaining to their students the basic concepts and principles of their fields, because it was held that this procedure would promote glib verbalism and rote memorization. It was felt that if students worked enough problems and were kept busy pouring reagents into a sufficient number of test tubes, they would somehow spontaneously discover in a meaningful way all of the important concepts and generalizations they needed to know in the fields they were studying.

---

†Numbers in brackets refer to the references at the end of this article.

Of course, one had to take pains to discourage students from rotely memorizing formulas, and then mechanically substituting for the general terms in these formulas the particular values of specified variables in given problems. This would naturally be no less rote than formal didactic exposition. Hence, in accordance with the new emphasis on "meaningful" problem solving, students ceased memorizing formulas, memorizing instead type problems. They learned how to work exemplars of all of the kinds of problems they were responsible for, and then rotely memorized both the form of each type and its solution. Thus equipped, it was comparatively easy to sort the problems with which they were confronted into their respective categories, and "spontaneously proceed to discover meaningful solutions"—provided, of course, that the teacher played fair and presented recognizable exemplars of the various types.

Similarly, as the terms "laboratory" and "scientific method" became sacrosanct in American high schools and universities, students were coerced into mimicking the externally conspicuous but inherently trivial aspects of scientific method. They wasted many valuable hours collecting empirical data which, at the very worst, belabored the obvious, and at the very best, helped them rediscover or exemplify principles which the teacher could have presented verbally and demonstrated visually in a matter of minutes. Actually, they learned precious little subject matter and even less scientific method from this procedure. The unsophisticated scientific mind is only confused by the natural complexities of raw, unsystematized empirical data, and learns much more from schematic models and diagrams; and following laboratory manuals in cookbook fashion, without adequate knowledge of the relevant methodological and substantive principles involved, confers about as much genuine appreciation of scientific method as putting on a white "lab" coat and doing a TV commercial for "Roll-Aids."

Partly as a result of the superstitious faith of educators in the magical efficacy of problem-solving and laboratory methods, we have produced in the past four decades millions of high school and college graduates who *never* had the foggiest notion of the meaning of a variable, of a function, of an exponent, of calculus, of molecular structure, or of electricity, but who have done all of the prescribed laboratory work, and have successfully solved an acceptable percentage of the required problems in differential and integral calculus, in logarithms, in molar and normal solutions, and in Ohm's Law. It is not at all uncommon, for example, to find students who have successfully completed a problem-solving course in plane geometry who believe that the descriptive adjective "plane" identifies the course as "ordinary" or "not fancy" rather than dealing with two-dimensional figures.

One basic lesson that some modern proponents of the discovery method have drawn from the educational disaster is that problem-solving per se is not conducive to meaningful discovery. Problem-solving can be just as deadening, just as formalistic, just as mechanical, just as passive, and just as rote as the worst form of verbal exposition. The type of learning outcomes that emerges is largely a function of the substance, the organization, and the spirit of the problem-solving experiences one provides. However, an equally important lesson which these same proponents of the discovery method refuse to draw, is that because of the educational logistics involved, even the best program of problem-solving experience is no substitute for a minimally necessary amount of appropriate didactic exposition. But this minimum will never be made available as long as we adhere to the standard university formula of devoting one hour of exposition to every four hours of laboratory work and paper-and-pencil problem-solving.

Developmental Limitations

A second psychological limitation of the discovery method is that on developmental grounds this technique is generally unnecessary and inappropriate for teaching subject-matter content, except when pupils are in the concrete stage of cognitive development. During the concrete stage, roughly covering the elementary-school period, children are restricted by their dependence on concrete-empirical experience to a semi-abstract, intuitive understanding of abstract propositions. Furthermore, even during these years, the act of discovery is not indispensable for intuitive understanding, and need not constitute a routine part of pedagogic technique. The only essential condition for learning relational concepts during this period is the ready availability of concrete-empirical experience. Thus, for teaching simple and relatively familiar new ideas, either verbal exposition accompanied by concrete-empirical props, or a semiautonomous type of discovery, accelerated by the judicious use of prompts and hints, is adequate enough. When the new ideas to be learned are more difficult and unfamiliar, however, it is quite conceivable that autonomous, inductive discovery enhances intuitive understanding. It presumably does this by bringing the student into more intimate contact both with the necessary concrete experience and with the actual operations of abstracting and generalizing from empirical data.

During the abstract stage of cognitive development, however, the psychological rationale for using discovery methods to teach subject-matter content is highly questionable. Students now form most new concepts and learn most new propositions by directly grasping higher-order relationships between abstractions. To do so meaningfully, they need no longer depend on current or recently prior concrete-empirical experience, and hence are able to by-pass completely the intuitive type of understanding reflective of such dependence. Through proper expository teaching they can proceed directly to a level of abstract understanding that is qualitatively superior to the intuitive level in terms of generality, clarity, precision, and explicitness. At this stage of development, therefore, it seems pointless to enhance intuitive understanding by using discovery techniques.

It is true, of course, that secondary-school and older students can also profit sometimes from the use of concrete-empirical props and from discovery techniques in learning subject-matter content on an intuitive basis. This is so because even generally mature students still tend to function at a relatively concrete level when confronted with a new subject-matter area in which they are as yet totally unsophisticated. But since abstract cognitive functioning in this new area is rapidly achieved with the attainment of a minimal degree of subject-matter sophistication, this approach to the teaching of course content need only be employed in the early stages of instruction.

Even when discovery techniques are helpful in teaching subject-matter content, we must realize that they involve a "contrived" type of discovery that is a far cry from the truly autonomous discovery activities of the research scholar and scientist. As a matter of fact, *pure* discovery techniques, as employed by scholars and scientists, could lead only to utter chaos in the classroom. Put a young physics student into a bathtub, and he is just as likely to concentrate on the soap bubbles and on the refraction of light as on the displacement principle that he is supposed to discover. In the UICSM* program, therefore, students are given a prearranged sequence of suitable exemplars, and from these they "spontaneously self-discover" the appropriate generalization. Under these conditions pupils are engaging in "true," autonomous discovery in the same sense that a detective independently "solves" a crime after a benevolent Providence kindly gathers together all of the clues and arranges them in the correct sequence.

Nevertheless, if we wish to be pedagogically realistic about discovery techniques, we must concede in advance that before students can "discover" concepts and generalizations reasonably efficiently, problems must be structured for them, and the necessary data and available procedures must be skillfully "arranged" by others, that is, simplified, selectively schematized, and sequentially organized in such a way as to make ultimate discovery almost inevitable. No research scholar or scientist has it quite this easy.

Subverbal Awareness and Discovery

In attempting to provide a sophisticated and systematic pedagogic rationale for the discovery method, Gertrude Hendrix has placed much emphasis on the importance of subverbal awareness. According to her, the achievement of subverbal awareness constitutes the essence of understanding, insight, transfer, and generalization, as well as the basic element of the discovery process; verbalization, on the other hand, is necessary only for the labeling and communication of subverbally achieved insights. Hendrix (1961) denies that verbal

> generalizing is the primary generator of transfer power. . . . As far as transfer power [is] concerned the whole thing [is] there as soon as the nonverbal awareness [dawns]. . . . The separation of discovery phenomena from the process of composing sentences which express those discoveries is the big new breakthrough in pedagogical theory [7, pp. 292, 290].

The "key to transfer," Hendrix [5, p. 200] states, is a "subverbal internal process—something which must happen to the organism before it has any knowledge to verbalize." Verbalization, she asserts further, is not only unnecessary for the generation and transfer of ideas and understanding, but is also positively harmful when used for these purposes. Language only enters the picture because of the need to attach a symbol or label to the emerging subverbal insight so that it can be recorded, verified, classified, and communicated to others; but the entire substance of the idea inheres in the subverbal insight itself. The resulting problem then, according to Hendrix [7, p. 292], becomes one of how to plan and execute teaching so that language can be used for these necessary

---

*University of Illinois Committee on School Mathematics.

secondary functions *"without* damage to the dynamic quality of the learning itself."

How plausible is this proposition? Let us grant at the outset that a subverbal type of awareness or insight exists, and that this type of insight is displayed by rats, monkeys, and chimpanzees in experimental learning situations, and by household pets, saddle horses, barnyard animals, wild beasts, children, and adults in a wide variety of everyday problem solving situations. But is it because of this type of insight that human beings have evolved a culture, and have achieved some progress in such fields as philosophy, chemistry, physics, biology, and mathematics, quite beyond anything yet approached by horses, chickens, or apes? Or is it because of the qualitatively superior transfer power of verbal or symbolic generalization?

The principal fallacy in Gertrude Hendrix's line of argument, in my opinion, lies in her failure to distinguish between the labeling and process functions of language in thought. She writes:

> We have been a long time realizing that subverbal awareness of a class, or a property, or a relation had to be in *some*one's mind before anyone could have thought of inventing a word for it anyway. In the natural order of events, the abstraction forms first, and *then* a name for it is invented [6, p. 335].

Now what Hendrix is referring to here is simply the labeling or naming function of language in thought. The choice of a particular arbitrary symbol to represent a *new* abstraction obviously comes *after* the *process* of abstraction, and is not organically related to it. But this is not the *only* role of language in the abstraction process, nor is it the *first* time that it is used in this process. Verbalization, I submit, does more than verbally gild the lily of subverbal insight; it does more than just attach a symbolic handle to an idea so that one can record, verify, classify, and communicate it more readily. It constitutes, rather, an integral part of the very process of abstraction. When an individual uses language to express an idea, he is not merely encoding subverbal insight into words. On the contrary, he is engaged in a process of generating a higher level of insight that transcends by far—in clarity, precision, generality and inclusiveness—the previously achieved stage of subverbal awareness.

The old philosophical notion that words merely mirror thought or clothe it in outer garments, is charmingly poetic but has little functional utility or explanatory value in the modern science of psycholinguistics. Even the seemingly simple act of making a choice of words in developing an idea, involves complex processes of categorization, differentiation, abstraction and generalization; the rejection of alternative possibilities; and the exclusion of less precise or over-inclusive meanings. All of these processes contribute to and help account for the qualitatively superior transfer power of symbolic generalization.

Although the transfer power of symbolic generalization operates at many different levels of complexity and sophistication, even the simplest level transcends the kind of transfer that can be achieved with subverbal insight. Consider, for example, the transfer power of the word "house," which most preschool children can use correctly. Obviously, before the child ever uses this word, he has some unverbalized notion of what a house is. But I submit that once he attains and can meaningfully use the verbal concept of "house," he possesses an emergent new idea that he never possessed before—an idea that is sharper, clearer, more precise, more inclusive, more transferable, and more manipulable for purposes of thinking and comprehension than its crude subverbal precursor. He can now talk about the idea of "house" in the abstract, devoid of all particularity, and can combine this idea with concepts of form, size, color, number, function, etc., to formulate relational propositions that could hitherto be formulated with only the greatest difficulty. That verbal concepts of this nature are more transferable and more manipulable than subverbal insights, is demonstrated by numerous experiments on the effects of verbalization on children's ability to solve transposition problems. Knowledge of underlying verbal principles also enhances the learning of relevant motor performance; and the availability of distinctive verbal responses facilitates rather than inhibits concept formation and conceptual transfer.

Not all ideas, however, are acquired quite as easily as the concept of house. As he enters school the child encounters other concepts of much greater abstractness and complexity, e.g., concepts of addition, multiplication, government, society, force, velocity, digestion, that transcend his immediate experience and language ability. Before he can hope to acquire a meaningful grasp of such abstractions directly, that is, through direct verbal exposition, he must first acquire a minimal level of sophistication in the particular subject-matter area, as well as graduate into the next higher level of intellectual development *i.e.,* the stage of formal logical operations. In the meantime he is limited to an intuitive, subverbal kind of understanding of these concepts; and even though convincing empirical evidence is still lacking, it is reasonable to suppose that preliminary acquisition and utilization of this subverbal level of insight both facilitates learning and transferability, and promotes the eventual emergence of *full* verbal understanding. (Gertrude Hendrix, of course, would say that *full* understanding was already attained in the subverbal phase, and that verbalization merely attaches words to subverbal insight.)

Now, assuming for the moment that Hendrix' experimental findings are valid, how can we explain the fact that immediate verbalization of newly acquired subverbal insight renders that insight less transferable than when verbalization is not attempted [5]? First, it seems likely that verbalization of nonverbal insight, before such insight is adequately consolidated by extensive use, may interfere with consolidation at this level, as well as encourage rote memorization of the ineptly stated verbal proposition. Even more important, however, is the likelihood that a verbally expressed idea—when ambiguous, unprecise, ineptly formulated, and only marginally competent—possesses less functional utility and transferability than the ordinarily more primitive and less transferable subverbal insight. This is particularly true in the case of children, because of their limited linguistic facility and their relative incompetence in formal propositional logic.

Drawing these various strands of argument together, what can we legitimately conclude at this point? First, verbalization does more than just encode subverbal insight into words. It is part of the very process of thought which makes possible a qualitatively higher level of understanding with greatly enhanced transfer power. Second, direct acquisition of ideas from verbally presented propositions, presupposes both that the learner has attained the stage of formal logical operations, and that he possess minimal sophistication in the particular subject matter in question. The typical elementary-school child, therefore, tends to be limited to an intuitive, subverbal awareness of difficult abstractions. The older, cognitively mature individual, however, who is also unsophisticated in a particular subject-matter area, is able to dispense with the subverbal phase of awareness rather quickly, i.e., as soon as he attains the necessary degree of sophistication; and once he attains it, he probably short-circuits the subverbal phase completely. Lastly, immediate verbalization of a nonverbal insight, when this latter insight is newly acquired and inadequately consolidated, probably decreases its transferability. This phenomenon can be explained by means of the general developmental principle, that an ordinarily higher and more efficient stake of development, while still embryonic and only marginally competent, is less functional than an ordinarily more primitive and less efficient phase of development. Running, for example, is eventually more efficient than creeping, but if a one-year-old infant had to run for his life, he would make better progress creeping.

Gertrude Hendrix, however, comes out with somewhat different and more sweeping conclusions from the same set of data. First, she regards non-verbal awareness as containing within itself the entire essence of an emerging idea, and insists that language merely adds a convenient symbolic handle to this idea. Second, she generalizes children's dependence on a preliminary subverbal stage of awareness, to all age levels, to all degrees of subject-matter sophistication, and to all levels of ideational difficulty. Actually, this subverbal stage is highly abbreviated, both for young children learning less difficult kinds of abstractions, and for older, cognitively mature individuals working in a particular subject-matter area in which they happen to be unsophisticated; and it is by-passed completely when this latter sophistication is attained. Finally, she interprets her experimental findings regarding the inhibitory effects of immediate verbalization on the transferability of subverbal insight, as providing empirical *proof* of her thesis that both the substance of an idea and the essential basis of its transfer power are present in their entirety as soon as nonverbal awareness emerges. In my opinion, these findings do nothing of the kind. They merely show that a relatively clear subverbal insight, even when only partially consolidated, is more functional and transferable than an ambiguous, inept, and marginally competent verbally expressed idea.

Unlike Gertrude Hendrix, therefore, I would conclude that secondary school and college students, who already possess a sound, meaningful grasp of the rudiments of a discipline like mathematics, can be taught this subject meaningfully and with maximal efficiency, through the method of verbal exposition, supplemented by appropriate problem-solving experience; and that the use of the discovery method in these circumstances is inordinately time-consuming, wasteful, and rarely warranted: Why then do discovery techniques seem to work so well in programs such as the one devised by the University of Illinois Committee on School Mathematics? For one thing, the students entering the program, being victims of conventional arithmetic teaching in the elementary schools, do *not* have a sound, meaningful grasp of the rudiments of mathematics, and have to be reeducated, so to speak, from scratch. For another, I have a very strong impression that as the program develops, the discovery element becomes progressively attenuated, until eventually it is accorded only token recognition. Lastly, stripped of its quite limited discovery aspects, the UICSM approach is a much more systematic, highly organized, self-consistent, carefully programmed, abstractly verbal system of verbal exposition than anything we have known to date in secondary-school mathematics. If it proves anything, the success of this program is a testimonial to the feasibility and value of a good program of didactic verbal exposition in secondary school mathematics, which program is taught by able and enthusiastic instructors, and in its early stages, makes judicious use of inductive and discovery techniques.

Time-Cost Considerations

From a practical standpoint it is impossible to consider the pedagogic feasibility of learning by discovery as a primary means of teaching subject-matter content without taking into account the inordinate time-cost involved. This disadvantage is not only applicable to the type of discovery where the learner is thrown entirely on his own resources, but also applies in lesser degree to the "contrived" or "arranged" type of discovery. Considerations of time-cost are particularly pertinent in view of our aforementioned developmental conclusions, that the discovery approach offers no indispensable learning advantages, except in the very limited case of the more difficult learning task when the learner is either in the concrete stage of cognitive development, or, if generally in the abstract stage, happens to lack minimal sophistication in a particular subject-matter field. Also, once students reach secondary school and university, the time-cost disadvantage can no longer be defended on the dual grounds that the time-consuming concrete-empirical aspects of learning must take place anyway, and that in any case elementary school pupils can't be expected to cover a great deal of subject matter. Thus, simply on a time-cost basis, if secondary-school and university students were obliged to discover for themselves every concept and principle in the syllabus, they would never get much beyond the rudiments of any discipline.

Some discovery enthusiasts (Bruner [2]; Suchman [9]) grudgingly admit that there is not sufficient time for pupils to discover everything they need to know in the various disciplines, and hence concede that there is also room for good expository teaching in the schools. In practice, however, this concession counts for little, because in the very next breath they claim the acquisition of actual knowledge is less important than the acquisition of ability to discover knowledge autonomously, and propose that pedagogy and the curriculum be reorganized accordingly. Thus, in spite of the formal bow they make to didactic exposition, it is clear that they regard the acquisition of problem-solving ability as more basic than the acquisition of subject matter. There is, after all, only so much time in a school day. If the school accepts as its principal function the development of discovery and inquiry skills, even with the best intention in the world, how much time could possibly remain for the teaching of subject-matter content?

Another disadvantage of using a discovery approach for the transmission of subject-matter content is the fact that children are notoriously subjective in their evaluation of external events, and tend to jump to conclusions, to generalize on the basis of limited experience, and to consider only one aspect of a problem at a time. These tendencies increase further the time-cost of discovery learning in the transmission of knowledge. Moreover, children tend to interpret empirical experience in the light of prevailing folklore conceptions that are at variance with modern scientific theories. Lastly, one might resonably ask how many students are sufficiently brilliant to discover everything they need to know. Most students of average ability can acquire a meaningful grasp of the theory of evolution and gravitation, but how many students can discover these ideas autonomously?

Training in the "Heuristics of Discovery"

Some advocates of the discovery method favor a type of guided practice in the "heuristics of discovery" that is reminiscent of the faculty psychology approach to improving overall critical thinking ability through instruction in the general principles of logic. For example, once the heuristics of discovery are mastered, they constitute, according to Bruner [3, p. 31], "a style of problem-solving or inquiry that serves for any kind of task one may encounter." In fact, one of the more fashionable movements in curriculum theory today is the attempt to enhance the critical thinking ability of pupils apart from any systematic consideration of subject-matter content. An entire course of study is pursued in which pupils perform or consider an unrelated series of experiments in depth, and then concentrate solely on the inquiry process itself rather than on this process as it is related to the acquisition of an organized body of knowledge.

One principal difficulty with this approach, apart from the fact that it fails to promote the orderly, sequential growth of knowledge is that critical thinking ability can only be enhanced within the context of a specific discipline. Grand strategies of discovery do not seem to be transferable across disciplinary lines—either when acquired within a given discipline, or when learned in a more general form apart from specific subject-matter content. This principle has been confirmed by countless studies, and is illustrated by the laughable errors of logic and judgment committed by distinguished scientists and scholars who wander outside their own disciplines. From a purely theoretical standpoint alone, it hardly seems plausible that a strategy of inquiry, which must necessarily be broad enough to be applicable to a wide range of disciplines and problems, can ever have, at the same time, sufficient particular relevance to be helpful in the solution of the specific problem at hand.

A second significant difficulty with this approach is that its proponents tend to confuse the goals of the scientist with the goals of the science student. They assert that these objectives are identical, and hence that students can learn science most effectively by enacting the role of junior scientist. The underlying rationale is that all intellectual activity regardless of level is of one piece, and that both creative scientists and elementary-school children rely heavily on intuitive thinking. Bruner [2, p. 14] is an eloquent spokesman for this point of view. According to him,

. . . intellectual activity anywhere is the same, whether at the frontier of knowledge or in a third-grade classroom. . . . The difference is in degree, not in kind. The schoolboy learning physics *is* a physicist, and it is easier for him to learn physics behaving like a physicist than doing something else.

It is also proposed that the ultimate goal of the Inquiry Training Program is for children to discover and formulate explanations which strive for the same universality and unification of concepts achieved by scientists (Suchman [9]).

First, I cannot agree that the goals of the research scientist and of the science student are identical. The scientist is engaged in a full-time search for new general or applied principles in his field. The student, on the other hand, is primarily engaged in an effort to learn the same basic subject matter in this field which the scientist had learned in his student days, and also to learn something of the method and spirit of scientific inquiry. Thus, while it makes perfectly good sense for the scientists to work full-time formulating and testing new hypotheses, it is quite indefensible, in my opinion, for the student to be doing the same thing—either for real, or in the sense of rediscovery. Most of the student's time should be taken up with appropriate expository learning, and the remainder devoted to sampling the flavor and techniques of scientific method. It is the scientist's business to formulate unifying explanatory principles in science. It is the student's business to learn these principles as meaningfully and critically as possible, and *then,* after his background is adequate, to try to improve on them if he can. If he is ever to discover, he must first learn; and he cannot learn adequately by pretending he is a junior scientist.

Second, there is, in my opinion, a world of difference between the intuitive thinking of elementary-school children and the intuitive thinking of scholars and scientists. The elementary school child thinks intuitively or subverbally about many complex, abstract problems, not because he is creative, but because this is the *best he can do* at his particular stage of intellectual development. The intuitive thinking of scientists, on the other hand, consists of tentative and roughly formulated "hunches" which are merely preparatory to more rigorous thought. Furthermore, although the hunches themselves are only make-shift approximations which are not very precisely stated, they presuppose both a high level of abstract verbal ability, as well as sophisticated knowledge of a particular discipline.

Development of Problem-Solving Ability as the Primary Goal of Education

In the realm of educational theory, if not in actual practice, exaggerated emphasis on problem-solving still continues to disturb the natural balance between the "transmission of the culture" and the problem-solving objectives of education. Enthusiastic proponents of the discovery method (e.g., Suchman [9]) still assert that

> more basic than the attainment of concepts is the ability to inquire and discover them autonomously. . . . The schools must have a new pedagogy with a new set of goals which subordinates retention to thinking. . . . Instead of devoting their efforts to storing information and recalling it on demand, they would be developing cognitive functions needed to seek out and organize information in a way that would be most productive of new concepts.

The development of problem-solving ability, is of course, a legitimate and significant educational objective in its own right. Hence it is highly defensible to utilize a certain proportion of classroom time in developing appreciation of and facility in the use of scientific methods of inquiry and of other empirical, inductive, and deductive problem-solving procedures. But this is a far cry from advocating that the enhancement of problem-solving ability is the *major* function of the school. To acquire facility in problem-solving and scientific method it is also unnecessary for learners to rediscover *every* principle in the syllabus. Since problem-solving ability is itself transferable, at least within a given subject-matter field, facility gained in independently formulating and applying one generalization is transferable to other problem areas in the same discipline. Furthermore, overemphasis on developing problem-solving ability would ultimately defeat its own ends. Because of its time-consuming proclivities, it would leave students with insufficient time in which to learn the content of a discipline; and hence, despite their adeptness at problem-solving they would be unable to solve simple problems involving the application of such content.

Discovery as a Unique Generator of Motivation and Self-Confidence

Bruner [3, 4] and other discovery enthusiasts (Hendrix [7]; Suchman [9]) perceive learning by discovery as a unique and unexcelled generator of self-confidence, of intellectual excitement, and of motivation for sustained problem-solving and creative thinking. It is undeniable that discovery techniques are valuable for acquiring desirable attitudes toward inquiry, as well as firm convictions about the existence and discoverability of orderliness in the universe. It is also reasonable to suppose that successful discovery experience enhances both these attitudes and convictions, and the individual's feeling of confidence in his own abilities. On the other hand, there is no reason to believe that discovery methods are unique or alone in their ability to effect these outcomes.

As every student who has been exposed to competent teaching knows, the skillful exposition of ideas can also generate considerable intellectual excitement and motivation for genuine inquiry, although admittedly not quite as much perhaps as discovery. Few physics students who learn the principle of displacement through didactic exposition will run half-naked through the streets shrieking, "Eureka." But then again, how many students of Archimedes' ability are enrolled in the typical physics or mathematics class? How comparable to the excitement of Archimedes' purely autonomous and original discovery, is the excitement generated by discovering a general formula for finding the number of diagonals in an *n*-sided polygon, after working problems one through nine in the textbook? And what happens to Archimedes Junior's motivation and self-confidence if, after seventeen immersions in the tub, he has merely succeeded in getting himself soaking wet?

Research Evidence

Despite their frequent espousal of discovery principles, the various curriculum reform projects have failed thus far to yield any research evidence in support of the discovery method. This is not to say that the evidence is negative, but rather that there just isn't any evidence, one way or the other—notwithstanding the fact that these projects are often cited in the "discovery" literature under the heading, "research shows." For one thing, the sponsors of some of these projects have not been particularly concerned about *proving* the superior efficacy of their programs, since they have been thoroughly convinced of this from the outset. Hence in many instances they have not even attempted to obtain comparable achievement test data from matched control groups. And only rarely has any effort been expended to prevent the operation of the crucial "Hawthorne Effect," that is, to make sure that evidence of superior achievement outcomes is attributable to the influence of the new pedagogic techniques or materials in question, rather than to the fact that the experimental group is the recipient of *some* form of conspicuous special attention; that *some*thing new and interesting is being tried; or that the teachers involved are especially competent, dedicated, and enthusiastic, and receive special training, attend expense-free conventions and summer institutes, and are assigned lighter teaching loads.

But even if the sponsors of the curriculum reform movements were all imbued with missionary research zeal, it would still be impossible to test the discovery hypothesis within the context of curriculum research. In the first place, a large number of other significant variables are necessarily operative in such programs. The UICSM program, for example, not only relies heavily on the principle of self-discovery of generalizations, but also on an inductive approach, on nonverbal awareness, on abundant empirical experience, on careful sequential programming and, above all, on precise, self-consistent unambiguous, and systematic verbal formulation of basic principles. To which variable, or to which combination of these variables and the "Hawthorne Effect" should the success of this program be attributed? Personally, for reasons enumerated earlier in this paper, I would nominate the factor of precise and systematic verbal formulation rather than the discovery variable. (Students enrolled in the UICSM program learn more mathematics in my opinion, *not* because they are required to discover generalizations *by themselves*, but because they have at their disposal a systematic body of organizing, explanatory, and integrative principles which are not part of the conventional course in secondary-school mathematics. These principles illuminate the subject for them and make it much more meaningful.)

Every Child a Creative and Critical Thinker

One of the currently fashionable educational doctrines giving support to the discovery method movement, is the notion that the school can make every child a creative thinker, and help him discover discontinuously new ideas and ways of looking at things. Creativity, it is alleged, is not the exclusive property of the rare genius among us, but a tender bud that resides in some measure within every child, requiring only the gentle, catalytic influence of sensitive, imaginative teaching to coax it into glorious bloom.

This idea rests on the following questionable assumptions: that one can be creative without necessarily being original; that all discovery activity, irrespective of orginality, is qualitatively of one piece—from Einstein's formulation of the theory of relativity to every infant's spontaneous discovery that objects continue to exist even when they are out of sight; that considering the multiplicity of abilities, every person stands a good chance, genically speaking, of being creative in at least one area; and that even if heredity is uncooperative, good teachers can take the place of missing genes.

Hohn's [8] use of the term "creativity" is typical of the prevailing tendency in "discovery" circles to "democratize" the meaning of this concept. A child behaves creatively in mathematics, according to Hohn, when he proposes alternative approaches, grasps concepts intuitively, or displays autonomy, flexibility, and freedom from perseverative rigidity in his discovery efforts. Now one can define words in any way one chooses, and hence can define creativity so that it means nothing more than "autonomous and flexible discovery." But if this is *all* one means, would it not save endless confusion if one used these particular words instead of a term which both connotatively and denotatively implies a rare form of originality?

As a matter of fact, the very same persons who use "creativity" in the more "democratic" sense of the term, also imply in other contexts that the encouragement of *true* creativity (that is, in the sense of original accomplishment) in *every* child, is one of the major functions of the school. This view is implicit in Bruner's [4] position that the school should help every child reach discontinuous realms of experience so that he can create his own interior culture. It is also implicit in the goal of the Inquiry Training Program, namely, that children should be trained to formulate the same kinds of unifying concepts in science which are produced by our most creative scientists (Suchman [9]).

How reasonable now is the goal of "teaching for creativity," that is, in the sense of singularly original achievement? A decent respect for the realities of the human condition would seem to indicate that the training possibilities with respect to creativity are severely limited. The school can obviously help in the realization of existing creative potentialities by providing opportunities for spontaneity, initiative, and individualized expression; by making room in the curriculum for tasks that are sufficiently challenging for pupils with creative gifts; and by rewarding creative achievement. But it cannot actualize potentialities for creativity if these potentialities do not exist in the first place. Hence it is totally unrealistic, in my opinion, to suppose that even the most ingenious kinds of teaching techniques we could devise could stimulate creative accomplishment in children of average endowment.

Even "teaching for *critical* thinking" and "teaching for problem-solving" are somewhat grandiose slogans, although obviously much more realistic than "teaching for creative thinking." To be sure, the critical thinking and problem-solving abilities of most pupils can be improved. But this is a far cry from saying that most pupils can be trained to become good critical thinkers and problem-solvers. Potentialities for developing high levels of these abilities are admittedly much less rare than corresponding potentialities for developing creativity. Nevertheless, there are no good reasons for believing that they are any commoner than potentialities for developing high general intelligence. Also, in my opinion,

. . . aptitude in problem-solving involves a much different pattern of abilities than those required for understanding and retaining abstract ideas. The ability to solve problems calls for qualities (for example, flexibility, resourcefulness, improvising skill, originality, problem sensitivity, venturesomeness) that are less generously distributed in the population of learners than the ability to comprehend verbally presented materials. Many of these qualities also cannot be taught effectively. Although appropriate pedagogic procedures can improve problem-solving ability, relatively few good problem-solvers can be trained in comparison with the number of persons who can acquire a meaningful grasp of various subject-matter fields [1, p. 23].

From the standpoint of enlightened educational policy in a democracy, therefore, it seems to me that the school should concentrate its major efforts on teaching both what is most important in terms of cultural survival and cultural progress, and what is most teachable to the majority of its clientele. As improved methods of teaching become available, most students will be able to master the basic intellectual skills as well as a reasonable portion of the more important subject-matter content of the major disciplines. Is it not more defensible to shoot for this realistic goal, which lies within our reach, than to focus on educational objectives that presuppose exceptional endowment and are impossible of fulfillment when applied to the generality of mankind? Would it not be more realistic to strive first to have each pupil respond meaningfully, actively, and critically to good expository teaching before we endeavor to make him a good critical thinker and problem-solver?

I am by no means proposing a uniform curriculum and pedagogy for all children irrespective of individual differences. By all means let us provide all feasible special opportunities and facilities for the exceptional child. But in so doing, let us not attempt to structure the learning environment of the *non*-exceptional child in terms of educational objectives and teaching methods that are appropriate for either one child in a hundred or for one child in a million.

References

1. Ausubel, D. P. "In defense of verbal learning," *Educational Theory,* XI (1961), 15-25.

2. Bruner, J. S. *The Process of Education.* Cambridge, Massachusetts: Harvard University Press, 1960.

3. Bruner, J. S. "The act of discovery," *Harvard Education Review,* XXXI (1961), 21-32.

4. Bruner, J. S. "After Dewey what?" *Saturday Review,* June 17, 1961, pp. 58-59; 76-78.

5. Hendrix, Gertrude, "A new clue to transfer of training," *Elementary School Journal,* XLVIII (1947), 197-208.

6. Hendrix, Gertrude, "Prerequisite to meaning," *The Mathematics Teacher,* XLIII (1960), 334-339.

7. Hendrix, Gertrude, "Learning by discovery," *The Mathematics Teacher,* LIV (1961), 290-299.

8. Hohn, F. E. "Teaching creativity in mathematics," *The Arithmetic Teacher,* VIII (1961), 102-106.

9. Suchman, J. R. "Inquiry training: building skills for autonomous discovery," *Merrill-Palmer Quarterly of Behavior and Development,* VII (1961), 148-169.

Exercises: Some Psychological and Educational Limitations of Learning by Discovery

1. Contrast the position of Ausubel with that of Davis. How can you reconcile their views? What are some of the strengths of verbal exposition? What are some of the weaknesses?

2. Ausubel complains that the way discovery advocates characterize exposition is simply setting up a strawman. Does Ausubel use the same strawman technique for discovery? Explain.

3. Examine the Ausubel—Hendrix controversy as Ausubel sees it. What is your position? Explain, citing specific points made by Hendrix and Ausubel.

4. What do you think are the positions of Bruner, Gagné, and Ausubel concerning discovery learning? How differently would each behave in teaching mathematics? Explain.

5. How do most adults learn? Is it as Bruner suggests or as Ausubel suggests? Give examples.

6. Does the nature of the subject matter influence one's choice of discovery or exposition? Explain. For which method is mathematics most suited? Why? How is mathematics most often taught? Why?

The Madison Project's
Approach to a Theory
of Instruction*

Robert B. Davis

In many fields today one finds on the one hand the "practical" man on the everyday firing line, and on the other hand, the "theoretical" man back in the laboratory. Presumably it suggests a condition of health for a subject when the two work harmoniously together. In the present remarks, I can speak only as the "practical" man, reporting on the everyday teaching and curriculum planning activities of the Madison Project, which is operated by teachers and mathematicians. We are greatly interested in a "theory of instruction," and we wish to contribute to it as much as possible from our admittedly untheoretical position. The theoretical work of Piaget and Bruner gives me considerable hope that our "practical" decision-making can be related to a broader theoretical perspective.[1]

I. The Madison Project

By way of background, let me explain that the Madison Project is one of the currently active "curriculum revision" projects, sponsored by the National Science Foundation, the U. S. Office of Education, and other agencies. We have basically set ourselves the task of seeking "the best experience with mathematics which can be provided for children at the pre-college level." As one result of this quest we have developed a program of supplementary work in algebra and co-ordinate geometry for grades 2-8 (*i.e.,* chronological ages 7 years through 13 years, approximately), we have developed a complete 9th grade mathematics course, and we are carrying on tentative exploration at the level of nursery school and kindergarten (chronological ages 3 years to 5 years, approximately).[2]

We have two aspects of direct interest to this conference. Firstly, we make films showing actual classroom lessons, and follow the same children (where possible) for many consecutive years.[3] You can watch on film one of our classes begin in grade 3 and gradually progress to grade 7; in another case, you can watch a group of students begin the study of Madison Project materials in grade 5 and continue through grade 9. much is revealed by these films, and we hope that many of you will wish to borrow them and to analyze them or comment upon them from your various points of view. I can think of nothing better than to accumulate

Robert B. Davis, "The Madison Project's Approach to a Theory of Instruction," *Journal of Research in Science Teaching,* 2 (1964): 146-162. Reprinted with permission.

*Financial support for the Madison Project is provided by the National Science Foundation, the Division of Co-operative Research of the U. S. Office of Education, and by other agencies.

essays or analyses of these films by, for example, cultural anthropologists like Jules Henry, psychologists like Jean Piaget, Jerome Bruner, Jerome Kagan, Henry Murray, Richard Suchman, and Richard de Charms, by appropriate mathematicians, logicians, teachers, psychoanalysts, and so on. I shall return to this later when we discuss the "description problem."

Secondly, the Project is, as I have mentioned, deeply interested in contributions to a "theory of instruction." A description of our everyday decision-making procedures—to the extent that we know them explicitly—constitutes the rest of this paper.

II. What Kinds of Mathematical Experiences Shall We Provide for our Students?

In general, there are two kinds of experiences which we provide for the children: experiences where children *do* something, and experiences where a "seminar" of children *discuss* something under the leadership of a teacher. Both kinds of experiences are so different from usual "mathematics lessons" that we have had to give them a distinctive name—*informal exploratory experiences*—in self-protection against unsympathetic observers who have told us "Why, there was no *teaching* in that lesson!"

It might be well to give one or two examples of each kind of "experience."

Angles and Rotations[4]

In these lessons, which we refer to as "experience" lessons, the children (at the 4th or 5th grade level in most cases) are shown pictures of angles drawn on the blackboard, asked to guess the measure of the angle (in degrees), and thereafter check their guesses by trying to measure the angle with a protractor, or with pie-shaped "units" (circular sectors of $10°$ central angle). They do "right-face," "about-face," and other turns with their own bodies (including turns through $30°$, $-30°$, $360°$, $720°$, and so forth), and rotate wheels through specified angles (of positive or negative measure). This kind of thing we refer to as "experience with angles." In a sense those observers are right who say "Why, there was no *teaching* in that lesson!" We believe there was, however, considerable learning. *The teacher has tried to bring the children into a direct face-to-face confrontation with the mathematics itself.* When such lessons work well, our teachers often have the feeling that they have somehow been able to step backward out of the way, and the child has been in direct communication with the mathematics.

Weights and Springs[5]

This lesson is mainly an "experience" lesson, but it sometimes turns suddenly into a seminar discussion lesson. Children (at the level of grade 4-9), perhaps working in small groups, hang weights on (A) a metal spring, and (B) a chain of rubber bands, recording at each step the weight on the spring (or rubber bands), and the amount that the spring stretches. We do *not* structure this by suggesting what weights they use, etc. We *do* suggest that they represent this data by a graph (but we do *not* tell them how to make the graph[6]). We also suggest that, if possible, they write an algebraic expression for the functional relationship shown on the graph.

This lesson becomes a "seminar" or "discussion" lesson if the children need to work together in a total group, under teacher guidance, in order to obtain the algebraic representation of the function, or if they wish to discuss sources of measurement error, the peculiar behavior of the rubber band, whether any apparent linearities are descriptive of the spring or are artifacts of the experimental procedure and of the measurement method, etc.

Matrices[7]

This lesson is usually a "seminar discussion" type lesson. Students who are already familiar with the structure of the rational numbers, and who know how to add and multiply matrices, are asked to explore the algebraic structure of the system of 2-by-2 matrices. (Grade level: 5 through 9, inclusive.) The point of the lesson might be stated as follows: in their previous work with the structure of the system of rational numbers, the children were getting experience in "exploring an unknown mathematical terrain." We now want to see how surefootedly they can go about the task of exploring *another* new mathematical terrain. The hope is that the children will know what kind of questions to ask, and what kind of answers to seek, as well as how to find these answers. (I shall return to this example later under the discussion of "shortcutting.") Obviously, where the children falter, the teacher tries to step in as unobtrusively as possible.

One way the teacher may do this is by making a suggestion that is, in fact, inappropriate. In the process of explaining to the teacher *why* the teacher's suggestion is inappropriate, the students are, of course, forced to peer more deeply into the mathematical structure itself. *Once again, the teacher has tried to remove himself from the role of middleman: he has tried to step out of the way and let the child look directly at the mathematical structure itself.*[8]

III. Criteria for Choosing Experiences

In selecting mathematical experiences to present to children, at the pre-college level, it is obvious that the danger of including inappropriate experiences is about as great as the danger of omitting valuable ones. The Project has developed a set of seven criteria for choosing appropriate experiences:

i. Adequate Previous "Readiness"

We try to make sure that, prior to the lesson in question, the children have had enough previous experience with essential ideas or techniques so that the desired new learning will be able to take place. I think our attention to prerequisites is unusually meticulous, in the sense that we seek a *very* careful breakdown of concepts into their simple "atomic" constituents.[9]

On the other hand, our pace is much more rapid than is customary in pre-college classes in the United States; we believe that our rapid pace is *not* the result of a neglect of readiness building, but is the result of a more optimistic expectation of student performance, a greater reliance upon the mathematical structure itself as a source of cognitive simplification ("reduction of cognitive strain")[10] and of motivation, and perhaps other similar factors.

ii. Relation to Fundamental Ideas

We do not wish to squander valuable momentum by a relatively unprofitable exploration of by-ways. Consequently, we make up a list of (what appear to us to be) fundamental concepts and techniques. This list includes: variable, function, Cartesian co-ordinates, open sentence, truth set, matrices, vectors, implication, contradiction, axioms and theorems, uniqueness, "mapping" or transformation, linearity, limit of a sequence, monotonicity, etc. *We require each "informal exploratory experience" to relate directly to these fundamental ideas.*

iii. The Student Must Have an Active Role

By this we mean to include activities such as problem-solving, arguing, criticizing, etc., as well as activities such as measuring, estimating, or performing an experiment. *We believe that many children fail in mathematics because they assume too passive a role.* In order to avoid this danger (which, in our view, is very great) we *almost never lecture,* and we make *very little use of required reading of routine material.* (Indeed, we probably make *too* little use of reading, but our fear of letting some students drop into a passive receptivity, with subsequent failure, is very great, and we tread carefully.)

iv. Concepts Must be Learned in Context

It appears to us that this is an important point. All of the paraphernalia of science or mathematics—concepts, equipment, data techniques, even attitudes and expectations—arise out of the act of tackling problems and arise out of inquiry. We want the concepts which the students form to arise in this same way. We believe this gives the ideas a different kind of meaning than they would have if they had sprung full-grown from the head of the teacher.

v. Interesting Patterns Must Lurk Under the Surface of Every Task

We want the students to form the habit of questioning even when there are no explicit external cues suggesting that they question. We wish them to be in the habit of asking: Did that really work? Can we extend it? *When* does it work? When would it fail? Is there a better way to do it?

In order to cause this "looking beyond the immediate problem" to become virtually habitual, we wish to give the students extensive practice in seeking underlying patterns. Our method for doing this is to attempt to arrange our material in such a way that there nearly always are some interesting patterns lurking just beneath the surface. They are usually not requisite for performance of the immediate task—but if one *does* observe them, they are interesting in themselves, they can usually be used to shorten greatly the effort required to perform the immediate task, they may suggest a more powerful attack that will solve harder problems, etc. One such example occurs when the immediate task is to practice the use of variables and the arithmetic of signed numbers—but this practice occurs in a context of quadratic equations, and sharp-eyed students can "discover" the important *coefficient rules for polynomial equations* lurking just beneath the surface. This discovery confers greatly added power to those who make it.

A second example occurs when we try to graph the truth set for linear equations. The immediate task can be handled by merely "plotting points," but a hugely extended power will be his who discovers the *patterns* of "slope" and "intercept."

vi. The Experiences Should Be Appropriate to the Age of the Child

This may seem to go without saying, but in ordinary education this precept seems honored mainly by non-compliance. We are finding, in our admittedly limited experience, that fifth graders (age about 10 years, chronologically) are "natural intellectuals," and can enjoy choosing a set of algebraic axioms and proving a variety of algebraic theorems from them. (This topic was formerly encountered in the latter years of college, or in graduate school.) By contrast, seventh and eighth graders are *not* "intellectuals"; it might come closer to say they are "engineers" at heart. For 7th and 8th graders, the usual school regime of sitting at desks, reading, writing, and reciting seems to ignore the basic nature of the child at this age; he wants to move around physically to do things, to explore, to take chances, to build things, and so on.* At this age we prefer to get the children out of their seats and, where possible, to get them out of the classroom and even to get them out of the school. We do vector problems by hanging twenty pound weights on yarn, and predicting whether the

---

*The importance of this question should not be overlooked. None of my reading in psychology, nor most of my contacts with psychologists, have attached much special importance to the 5th-6th-7th-8th-grade developmental pattern. Yet for virtually every one of the "new curriculum" projects in mathematics and science which deal experimentally with this age range, this is the single decisive, elusive, and discouraging phenomenon. The 5th grader is very good at mathematics and science. This *same* child, at grade 6, begins to perform less well. In grades 7 and 8 he is usually a total loss—he will perform routine tasks to a mediocre standard, but in situations calling for great creativity he usually creates chaos. Since encountering this catastrophe, we have accumulated about 20 alternative explanations, from psychoanalysts, teachers, physiologists, and parents. They include: (i) the sex theory: a sexual revolution and awakening is occurring, and all else is secondary; (ii) the energy theory: the child's energy is tied up in physical growth (which occurs rapidly at this age); (iii) the metabolic theory: the child's metabolic rate shifts, and it takes him several years to adjust his behavior to the new metabolic rate; (iv) the "noise" theory: everything we teach is wrong; by grade 6 the child has been in school long enough to accumulate so much misinformation that he is lost; (v) the "nobody loves junior high school" theory: junior highs get the almost-discarded buildings, the almost-discarded teachers, and the almost-discarded objectives and methods; (vi) the peer-group theory: the 5th grader loves adults; the 7th grader knows better, and believes his contemporaries, unpromising though they appear, are in the long run a better bet; (vii) the neo-Pareto theory: every generation has to take over from its elders, and grade 7 is the place to start; (viii) the "finding yourself" theory; (ix) the "finding reality" theory: 5th graders are remote from reality and allow themselves an interest in abstract things; a 7th grader is becoming sensitive to power, and so he demands more "practical" employment; (x) the "poor self-concept of the junior high teacher" theory: high school teachers really want to be college teachers, and commit the folly of frustrated emulation; junior high teachers imitate the imitators, and that's even worse. Many other theories have been proposed.

yarn will hold or will break; we do graphical differentiation and rate-of-change problems in the context of velocity and acceleration, using actual automobiles in the school driveway; we determine the height of the school flagpole by similar triangles; and so on.[1]

We do not know whether we are on the right track or not, but to our amazement we find no established and well-accepted theory to help decide this problem: what *kind* of school experience should the 7th and 8th grader have? In our own clinical interviews with children of this age, we find they greatly prefer "moving around" subjects (like gym, dancing, shop, art, music, home economics, science laboratory, etc.) over sedentary subjects (such as Latin, English, mathematics, and social studies). We need to understand this far more than we do; in the meantime, we are asking ourselves if mathematics, social studies, etc., *need* to be sedentary subjects at this grade level. (Notice that *traditional* dogma emphasizes physical activity for *younger* children—some of whom, in our experience (*e.g.,* fifth graders) do not especially require it—but places *less* emphasis upon physical movement for the older 7th and 8th graders. Obviously the over-all school program must be considered as a causal factor, also; the younger children, in most schools, may "get enough" chance for physical movement, whereas in junior high school they may not.)

vii. The Sequence of "Informal Exploratory Experience" Must Seem to "Add Up" to Something Worthwhile

By this we mean that the teacher (or other observer) must feel that the lesson, the day, the week, the year have each made their proper contribution to the child's growth toward mathematical maturity and sophistication.

In selecting "informal exploratory experiences," we keep in mind all seven criteria listed above.

IV. The Flexibly-Programed Discussion Sequence

In our "experience" lessons we structure the situation as little as possible. (In practice, we sometimes structure it too little and sometimes too much.)[12] In our "seminar discussion" lessons there is at once an appearance of a relatively highly structured situation, and at the same time an appearance of great flexibility. After wondering about this seeming paradox for some time, we have come to believe that the "good" Madison Project teacher possesses in his head the ability to construct suitably designed "branching programs" at a moment's notice.

Let me give an example. The film, *A Lesson with Second Graders,* shows a sequence where the teacher poses the problem

$$^+5 + \square = {}^+3$$

and asks the students what number can be substituted to produce a true statement. A girl named Charlotte responds immediately with "positive two." At this point the teacher's internal program-constructing facility devises a suitable "error-correction loop." He asks how much is

$$^+5 + {}^+2 = ?$$

Charlotte thoughtfully (or hesitantly) volunteers the answer "positive seven . . ."

The teacher now asks: what can we write in the □ to make a true statement from the open sentence

$$^+5 + □ = {}^+3$$

and Charlotte correctly answers "negative two."

I do not mean that every Madison Project teacher would *describe* his method of operation in the terms I have used here, but I believe that good Madison Project teachers do *behave* this way.

Indeed, the study of the properties of the "program sequence" which teachers use, perhaps together with a study of the cues which determine teacher decisions, may represent one of the most powerful points of attack in developing a theoretical understanding of "Madison Project teaching."

V. Reinforcement Schedules

Psychologists* observing Madison Project sessions have repeatedly emphasized the quite unusual use (or non-use) of reinforcement schedules in Project classes. We should admit at the outset that we use the ordinary "rewards" such as praise and affectionate warmth, etc. in securing reasonable *social* behavior. *We try never, however, to use a teacher-imposed external reinforcement schedule to determine what a child thinks, how he answers a question, or how he attacks a problem.* (This last may be a mild over-statement; perhaps the proper description should be: well, hardly ever.)

We try to use two forms of reinforcement only: first *intrinsic* rewards derived from solving a problem, from the reduction of cognitive strain which follows upon the discovery of an important concept or relationship, from the gratification of experimental verification of a prior theoretical prediction, etc.,† and, second, the reward that comes from being able to *tell* your classmates or your teacher about what you have just discovered or have just accomplished.

As a consequence of our inclusion of this second motivating factor, there is a good deal of communication going on in Project classrooms. In particular, we have never really developed the highly individualized instruction represented by "each student alone," as this is seen in graduate chemistry laboratories, or in the use of some of Z. P. Dienes' arithmetical materials at the elementary school level (as in Leicestershire, England, or at Shady Hill School in Cambridge, Massachusetts).

Personally, I suspect the Project may have developed too deep an attachment to "seminar-type discussion" lessons, and has, in consequence, failed to develop enough "informal exploratory

---

*For example, Dr. Richard Anderson of the University of Illinois, and Dr. Carl Pitts of Webster College.

†Kohler, for example, speaks of certain situations—such as an expensive cup and saucer teetering precariously on the very edge of a table—that stresses "demand quality." We feel that virtually every good lesson should be based upon tasks, situations, problems, questions, seeming contradictions, etc., that possess "demand quality." *In this* case we believe that we can entirely eliminate frill—that is to say, we hope to eliminate *all* of those tasks for which the child sees no reason other than to please the teacher.[11]

experiences" in a form where each child works alone, or where the children co-operate in quite small groups (with perhaps three children per group). We are now trying to remedy this, but progress is slow. Is it possible that this slow progress results from the fact that "each child working alone" is *not* an entirely natural and satisfactory situation? Perhaps the reinforcement of showing the whole class how clever you are really is virtually essential. The tendency of mature scientists to announce their results, often at considerable length, should not be overlooked in this connection.

VI. Autonomous Decision Procedures

We believe that the child should, everywhere possible, have a method for telling whether an answer is right or wrong that is independent of the teacher and independent of the textbook. For physical scientists the laboratory ostensibly fills this need. For mature mathematicians, logic ostensibly fills this need.[13] We have tried to fill this need in the earlier grades by using *counting* to verify work in arithmetic, and in later grades we try to provide multiple methods for solving problems as one way to decide correctness (that is, by the agreement or disagreement of results obtained by different methods). As a second "autonomous decision procedure" we try to provide models, as in the case of "postman stories" for the arithmetic of signed numbers.[14]

It appears to us that removing the "correctness" of mathematics from the authority of the teacher greatly increases student motivation. In the words of Jerrold Zacharias, "Science is a game played against nature; it is not a game played against the teacher."

VII. Degree of Autonomous Control

We also believe that the more freedom we can give the children, the more easily we can maintain a *really* high level of motivation.[15] Since we follow the same children as long as five years, this is a matter of some importance to us.

The nature of "freedom" has puzzled mankind for a long time, and I do not claim that we understand it. We can, however, say that when a child feels that a task is artificial, capriciously (or thoughtlessly) imposed by the teacher, he does not usually regard it as a serious challenge. Where (as usually, in our work) the task is determined by the teacher, the greater the extent to which a child is free to define his own method of attack, to define the "boundary conditions," and to define for himself what shall constitute an acceptable answer, the more he is inclined to take the whole matter seriously.

I do not claim that we understand "freedom," but we can recognize situations where the child enjoys but a very low degree of autonomous control, and in such situations we do not believe that very much good learning takes place. Even more apparent, in such situations the level of *continuing* motivation is usually not very high for most students.

VIII. Assimilation and Accommodation

It might be well to describe our understanding of some of Professor

Jean Piaget's theoretical remarks, specifically those dealing with the gradual modification of existing cognitive structure.

In the first place, an orientation based upon the notion of the *gradual modification of the individual's internal cognitive structure* appears to us as highly appropriate for studying the learning of mathematics. This is the kind of task with which the math teacher and the math learner is confronted. Indeed, at one time I thought that the two basic tasks of the mathematics teacher were to show important differences between matters which the student erroneously considered to be the same, and to point out similarities between situations that the student had failed to relate to one another; that is to say, to sever notions incorrectly related and to relate notions incorrectly separated.[16] If I now believe that the teacher's task is more complex than this, I do not for a moment doubt that both teacher and learner begin with the learner's "initial" cognitive structure, and seek to re-shape this into a "more sophisticated" cognitive structure. *This, again, is a powerful point of attack for studying what is involved in teaching and learning mathematics.*

We, as teachers, often think of *assimilation* and *accommodation* in terms of the task of learning to find your way around a strange city. At first there is so little cognitive structure that you cannot make sense out of directions, observations, etc. Presently one builds up such basic concepts as a knowledge of the principle streets and main landmarks. One can either extend the picture by introducing additional detail, or when "paradoxes" are encountered, modify the picture by removing major errors that, previously unnoticed, have suddenly become important.

This view of learning of mathematics has many direct implications for the classroom: for one thing, it encourages the use of "readiness building" and preliminary "unstructured exploration," in order to allow the child to build *some* basic relevant cognitive structure which more systematic instruction can then seek to modify. Again, this picture is consonant with the fact that everything that any of us knows is *wrong*. You cannot, from memory, even sketch a floorplan of a home you have lived in for years, without committing many "errors" and leaving plenty of room for further improvement. Hence it is foolish to say that "we shall see that the child learns everything correctly from the very beginning." It is equally foolish to say that "we shall never allow a child to leave a class with a misconception in his mind." Every idea of every one of us on every subject is wrong—partly wrong, that is. We learn by successive approximations, and there *is* no final and absolutely perfect "ultimate version" in any of our minds. We are wrong, but we can learn; having learned, we shall still be wrong, but less so; and, after that, we can still develop a yet more accurate cognitive representation, within our minds, for the various structures that exist independently of our minds.

This may seem obvious. Nonetheless, we encounter time and again the teacher (*not* a Madison Project Teacher!) who says that her students must "get things exactly right from the very beginning, so as not to learn any wrong ideas," and the question (asked, for example, of Richard Suchman, after a demonstration class in Los Angeles recently) "Would you allow that student to leave school that day with a wrong idea?"

The anguish of living with a complex reality may leave us sorely tempted to seek absolute and over-simplified revisions, which are not to be subjected to further modification, but this is not the way to learn a creative approach to modern science and mathematics, nor is it the way to run a democracy. Growth by successive approximations is the most we can hope to achieve.*

There is an important point, however.

When we say we cannot protect the child from "wrong" ideas, have we removed all possible *values* from a theory of instruction?

The answer, of course, is that we have not. We can think of a sequence of cognitive structures,

$$\ldots C_m, C_{m+1}, C_{m+2}, \ldots$$

where each is "more suitable" or "more sophisticated" than any that came before. Each picture is an imperfect model of "reality," *but it has other attributes upon which we can pass judgment.* If we cannot protect the child from "error," we *can* (and must) pass judgments on such matters as:

> Is this particular cognitive structure, C_p, suitable for the *assimilation* of the ideas with which we are presently working?
>
> Is this particular cognitive structure, C_p, a suitable one from which we can ultimately get to a more sophisticated structure, C_{p+1}?
>
> Are the emotional, social, and cognitive aspects of our classroom such that the child will easily move from one cognitive structure, C_p, to various more suitable ones, $C_{p+1}, C_{p+2}, \ldots$?
>
> At what point is it desirable for the child to become aware of some of the limitations of a given structure, C_p? When should he develop the new structure C_{p+1}? What should we, as the teacher, do in order to play midwife to the birth of structure C_{p+1}?

The ex-student who graduates from our program of educational experiences should, among other desirable attributes, be adept in discarding one cognitive structure and replacing it with a more adequate new one, and he should have the wisdom to know when this is advisable.†

Incidentally, this notion of a sequence of cognitive structures helps illuminate an important remark of David Page, that every child always gives the right answer to every question, as seen from his point of view. One can re-word this to say that the child must map the question into his cognitive structure as best he can, he must seek within his cognitive structure for answers, and, within these constraints, his answer is probably the most appropriate that can be given.

IX. The Danger of "Short-Cutting"

There is a peculiar phenomenon which we do not understand. Put briefly, it is this: if one states a specific set of really explicit objectives for an educational experience, this list seems always to be *significantly* incomplete: it is always possible to meet all of the stated requirements, *without actually achieving what was really desired.**

---

*"Oh, Lord," Herman Melville once wrote, "shall we never be done with growing?"

†In the words of John W. Gardner, "The ultimate goal of the educational system is to shift to the individual the burden of pursuing his own education." (*Science*, February 14, 1964).

*The situation suggests Mark Twain's comment on his wife's use of profanity: "She knows all the words, but she can't quite get the tune."

Why this should be so, or to what extent it *is* so, we do not feel we understand. However, it causes us to be wary of any approach to education which presumes an explicit *a priori* listing of objectives. At present we feel that any approach which depends upon a specific listing of objectives, however rational this may *seem*, is in fact an open invitation to somehow losing sight of the subtle, but *unstated* values *which are the real point of it all.*[17]

Here are some examples: a large vocabulary may mark the well-educated man; but we can train people to use a large vocabulary while somehow omitting many essentials of a good education.

Again, the enthusiasm of a man with an idea may often be a measure of the worth of the idea or of the nature of the man. Yet a certain popular program for "making a good appearance" advises students to show enthusiasm *even when they do not feel it, and even when they do not really have very much of an idea in mind at all.*

We regard this kind of "short-cutting," which achieves the appearance without regard for the reality, as pernicious in the extreme. But one invites precisely this whenever one lists in advance those appearances which will constitute "success." Indeed, as we shall see in the next section, traditional ninth-grade algebra puts great emphasis on getting students to write formulas that appear correct, *even though the students did not know what the formulas meant.*†

X. Mathematics Teaching, U. S. A., 1964

I have made some remarks on the kind of school experiences that the Madison Project seeks to achieve, and on some that it seeks to avoid. Let me now describe what we have seen in our role as observers, sitting in the back of non-Madison Project classes. This is what happens in "traditional" teaching in the U. S. A. in 1964:[18]

In the first place, the student encounters a slow-moving sequence of pedestrian tasks that, motivationally, require him merely to do what he is told, when he is told to, in the way that he is told to do it. From an intellectual or cognitive point of view he may encounter difficulties, but they are not the difficulties endemic to the subject matter; they are the difficulties engendered by obscure communications between teacher and student and by uncertainties or vagueness within the teacher's own mind. I could give many anecdotes to show that the *pace* is really incredibly slow. Here is one: a 9th grade class of rather bright children spent two consecutive periods, a total of nearly two hours, responding to questions such as

$$x^2 \times x^3 = ?$$
$$p^{10} \times p^7 = ?$$

by answering, respectively, "x^5" and "p^{17}."

An hour and a half later we find these children still working at this same task:

---

†A major automobile manufacturer is said to have a department concerned only with the sound when a door is slammed, presumably on the theory that the sound of the door slamming is used by many people to judge the "solidity" of the car.

Teacher: $x^5 \times x^3 = ?$

Class: x^8

This is apparently the neo-Pavlovian method for teaching algebraic notation to dogs. It should be noticed that no algebraic *concepts* are involved, merely typesetters' symbols that seemingly denote nothing. Moreover, if these students show up in college, they will be so conditioned that the stimulus

$$x^2 + x^3$$

will probably evoke the response

$$x^5$$

Whose fault is this? Is this *really* the human use of human children? Is this *really* what psychologists teach teachers to believe? The teacher in question, though surely misguided, was conscientious and legally certified.

Second, as the preceding example shows, teachers never think to use *the internal structure of the mathematics itself* as a source of motivation, a safeguard against confusion, a protection against forgetting, an alternative to unmindful drill. As Warwick Sawyer has put it, "the poor teacher asks the child: 'Don't you remember the rule?,' whereas the good teacher asks 'don't you see the pattern?' "

Having opted against organizing learning in terms of the internal structure of the subject to be learned, what can the teacher put in its place? Typically, she finds four things: *imitation* (in the "rule-example-drill" trilogy), the use of *external reinforcement schedules* ("There's no *reason!* It's company *policy.*"), neo-Pavlovian *drill* (as in the case above), and *cultural determinism*. This last point is of some interest. We can take any present-day citizen of the United States, and get him to answer certain questions "correctly," *even though he has no idea what he is talking about,* simply by using the fact that he is a present-day citizen of the United States. This is commonly done and is evil.

Here is an example. Within many number systems, every element has an additive inverse. The additive inverse is *itself* an element of the system, and so *it* has an additive inverse. Can you discuss this last-mentioned element?

Perhaps not. Indeed, on the bare facts given, several possible answers might be correct and none is obvious.

The problem can be re-stated. In place of the mathematically accurate phrase "additive inverse," substitute the suggestive word "opposite." It is now *obvious* to the student that the opposite of the opposite is the element you started with! Obvious, but unfortunately not universally true! This kind of culturally-determined answer creates a spurious structure which is not the genuine structure of the mathematics itself, but an *ersatz* structure "borrowed" from the culture. The student who learns this comes to see things with an out-of-focus fuzziness that makes it very hard for him to see the *real* structure of the mathematics, which is what he was supposed to have been learning.

As a second example, teachers sometimes say "negative one times negative one makes positive one, because two negatives make a positive." Some weaknesses of this argument are at once apparent: Why negative *one,* why not negative *two?* Does the analogy with English imply that, when Professor Piaget is in the United States, he should write

$$-1 \times -1 = +1$$

but when he returns to his French-speaking home, where two negatives work quite differently, he should write

$$-1 \times -1 = -1 ?$$

The *worst* part of this kind of argument, though, is that, *by substituting a spurious "culturally-determined" structure, it badly obscures the student's view of the genuine structure of mathematics.*

A consequence of "teaching" mathematics so that it is the moral equivalent of memorizing the telephone directory is that students know not what their words could mean and should mean, and they know not that they know not. We have called this the "superficial verbal problem." (Professor Raphael Salem, of M.I.T., once remarked that any M.I.T. freshman could recite "the derivative of log x is 1 over x," but the student did not know what a "log" is, and didn't know what a "derivative" is.)*

XI. The Teacher Needs to Listen to the Child

It may seem superfluous to mention that the teacher needs to listen to the child as carefully as she can. Nonetheless, and this may be symptomatic of the teacher's view of the process of learning, many teachers do not, and many fine and creative responses of students are rejected as "wrong" because they do not conform to the *a priori* expectations of the teacher.

Indeed, *listening to the child's suggestions* appears to us to be one of the cornerstones upon which "new mathematics" is built. Do not require the child to read your mind; do not require him to "do it your way"—show a simple respect for the child's intellectual, analytical, and problem-solving autonomy and you will discover that he is a much cleverer child than you had ever imagined!

How *can* the child improve his own personal internal cognitive structure when it is only the *teacher's* cognitive structure that is ever discussed?[19]

XII. The "Description" Problem

In work with children and using our "informal exploratory experiences," we sometimes get excellent results. We have had fifth graders conjecture special cases of the binomial theorem and prove them from a set of axioms selected by the children themselves. We have had

---

*One can distinguish "culturally-determined" responses from "mathematically determined" responses by seeking examples where the two responses would be different. For example, cf. the question:

A set which is not open is _____.

The "culturally-determined" response is surely "closed." This, however, is *mathematically* incorrect, since *mathematical* sets may be closed, open, neither, or both!

sixth graders use an isomorphism between rational numbers and a subset of the set of 2-by-2 matrices in order to solve the equation

$$x^2 = -4$$

and to introduce complex numbers. We have had 8th graders work out a theory of infinite sequences and use this as a foundation for the introduction of irrational numbers. In some classes interest of virtually *all* students runs high, and *continues to run high for five consecutive years.* In some situations, students voluntarily attend classes before school in the morning, or on Saturdays.

But in the case of other classes, this fails to happen. Student interest does *not* become great, and what there is does not maintain itself at a very high level. Student participation may be poor, and "brilliant student achievements" may not occur.

What makes the difference?

The response does not seem to be adequately explained in terms of I.Q. differences. We have, at various times, wondered about many possible explanations including these:

(*i*) Recalling our extensive (perhaps excessive) use of "seminar-type" informal discussion, do we get better results with *homogeneous* classes where a wise guidance counselor hovers in the wing to maintain a "compatible" group of children in the seminar?

(*ii*) Is it a question of which children in the class set the dominant tone for the class?

(*iii*) Is it a question of the kind of status fights that are going on among the children?

(*iv*) Since much of the Madison Project material is taught by a "visiting specialist teacher," is it a question of the effectiveness of the *classroom teacher* in welding the children into a cohesive, co-operative team that can work together, much like a good basketball team, where "setting up" intellectual shots contributes to "making baskets?"

(*v*) Since Madison Project teaching lays considerable stress upon freedom, autonomy, and responsibility, is it a question of how consistent this orientation is with that of the regular classroom teacher, of the school, and of the community?

(*vi*) Recalling again that the "informal exploratory experiences" are usually supervised by a "visiting specialist teacher," is the success or failure determined by the degree of "moral support" provided *by the regular classroom teacher* (or its opposite—by the deliberate attrition effected by the regular classroom teacher)?

(*vii*) Does success or failure depend upon parental attitudes? (This, and item (*vi*), include also the status or prestige assigned to the visiting specialist teacher, which is sometimes very high and sometimes is not.)

In answering questions of this sort, we encounter at least three obstacles:

(*a*) We are not interested in individual variables operating *in vacuo;* the question, for example, of "teacher attitude" must be operative in an actual Madison Project situation, where the other parameters fall within a "typical" range of values;

(*b*) We cannot alter school situations at our will; in the words of Bruner, it is not true that "the universe is spread before one, and one has freedom of choice as to what one will take as an instance for testing." Instead, the plight of the experimenter is "that he must make sense of what happens to come along, to find the significant grouping in the flow of events to which he is exposed, and over which he has only partial control."

Methodology for such situations does, of course, exist, but it requires the creation of a large situation for study, that in turn requires very extensive teacher training, the development of a large repertoire of "informal exploratory experiences," the recruiting of a large number of co-operating schools, and so on.

(c) Especially fascinating is the difficulty in finding behavioral scientists who are willing to look at *what goes on in the school and in the classroom.* Each has his own personal professional specialty, and, usually disdaining an over-all view of what is going on, he prefers to begin work at once in terms of his own specialty: motivation, or anxiety levels, or peer-group status structure, or clarity of cognitive aspects of discussion, or problems in the measurement of divergent thinking, or whatever. Just as it has been difficult to find mathematicians with a deep interest in education, it has also been difficult to find behavioral scientists who try to locate the forest first, before they try to focus on individual trees. This is admittedly a common problem in academic life today, and has parallels everywhere, for example in medicine, where most physicians will focus on certain aspects without seeing the patient as a whole.

What we seek, then, is a general description of what goes on in the classroom and in the school, from which we can begin to identify those variables which appear to be most decisive in determining success or failure, in the long run, for our program of "informal exploratory experiences."

XIII. The Measurement of Dependent Variables

What happens to children as a result of their participation in an extended program of "informal exploratory experiences" in mathematics? In trying to measure the outcome in terms of explicit dependent variables, we have encountered many obstacles, including these:

(i) In current parlance, we are concerned with "divergent thinking." Actually, I believe the "divergent-convergent" dichotomy is a mis-statement of the real problem, but we are, in any event, concerned with *the way the child explores on his own, his original "creative" ideas, and other responses which he will not necessarily produce in response to external cues.*

(ii) The more noteworthy student achievements are unlikely events anyhow and are easily masked by statistical "noise" in testing procedures.

(iii) The child's *attitude* is surely a most important aspect of the outcome.

(iv) Especially difficult is the fact that in education, as in quantum physics and brain surgery, *the act of measuring requires a significant alteration of the situation.* In an informal situation where teacher and students are colleagues and co-workers, with great mutual respect and autonomy, *can the teacher afford to seem to pass judgment on the student?* Can the teacher poke and probe without finding that the student is cognitively ticklish?

XIV. What Is "Discovery?"

It should be clear that one of the key questions before us today is the question of what we mean by "learning by discovery," and what good is it?

I feel sure that disagreement over the nature and value of discovery is rooted mainly in disagreement over *values*. If one thinks of arithmetic as a routine skill which the student should master—if this view is *uppermost* in your mind—then you will probably find no advantage in teaching by "discovery."[20]

The pre-school child (as witness my own two-year old daughter) lives a waking day that is an unending orgy of exploration and discovery. When the child enters school we try to discipline this exploratory propensity. There is abundant evidence that the desire to explore—at least, as Alan Waterman says, in academic situations—withers and just about dies. The decline of inquisitiveness in academic situations is perhaps not too pronounced by grade 5; by grade 6 it will usually make an unmistakable appearance; and by grades 7, 8, or 9 you can no longer feel any pulse.

Here is an actual example: ask 5th graders what they can write in the "□" to produce a true statement from the open sentence

$$\square \times \square = 2$$

They try 1; it is too small: $1 \times 1 = 1 < 2$
They try 2; it is too large: $2 \times 2 = 4 > 2$
They try 1.5; it is too large: $1.5 \times 1.5 = 2.25 > 2$
They try 1.4; it is too small: $1.4 \times 1.4 = 1.96 < 2$

Now they are off to the races, and they produce a great spate of suggestions, approximations, and relevant questions.

Propose the same question to older children; the older the child, the less creative and enthusiastic the response, until finally by the college freshman year, students very frequently respond by saying: "I don't think we had that in our high school." That is to say: *if nobody ever told me, then I cannot possibly know!*

Is this a triumph or a failure of education?

To clarify, if possible, our interest in "discovery," let me list some of the objectives of Madison Project teaching:[21]

1. "Cognitive" or "Mathematical" Objectives:

(*i*) the ability to discover pattern in abstract situations;

(*ii*) the ability (or propensity) to use independent creative explorations to extend "open-ended" mathematical situations;

(*iii*) the possession of a suitable set of mental symbols that serve to picture mathematical situations in a pseudo-geometrical, pseudo-isomorphic fashion, somewhat as described by the psychologist Tolman[22] and the mathematician George Polya;

(*iv*) a good understanding of basic mathematical concepts (such as variable, function, isomorphism, linearity, etc.) and of their inter-relations;

(*v*) reasonable mastery of important techniques;

(*vi*) knowledge of mathematical facts.

2. More General Objectives:

(*i*) a *belief* that *mathematics is discoverable;*
(*ii*) a realistic assessment of one's own ability to discover mathematics;
(*iii*) an "emotional" recognition (or "acceptance") of the open-endedness of mathematics;
(*iv*) honest personal self-critical ability;
(*v*) a personal commitment to the value of abstract rational analysis;
(*vi*) recognition of the valuable role of "educated intuition";
(*vii*) a feeling that mathematics is "fun" or "exciting" or "challenging" or "rewarding" or "worthwhile."

Actually, there is another important objective. We want the child to know who he is in relation to the human cultural past. By developing mathematics through *discovery* and through *student initiative, we have brought history right into the classroom!* The students have struggled with

$$x^2 = 2,$$

have been stymied, have tried various tangents and flank-attacks, and have finally witnessed a major historical breakthrough when some student proposed the adoption of an axiom that *every bounded monotonic sequence converges.*

These students really know what a "historical breakthrough" means; *they have lived through many, right in their own classroom.*

XV. Symbol Systems and Language

Professor Bruner has placed considerable stress on the role of language in cognitive functioning. If one means *the English language,* this is *not* terribly useful *in mathematics.* The symbols do not fit the ideas. If, however, one thinks either in terms of Tolman-esque intuitional symbols of a pseudo-geometrical nature, or else of appropriate mathematical notations, then the greater power conferred by these two (quite different) kinds of symbol systems is dramatic. We are overwhelmed at the greater power our students (in grades 5-9) have even from the simple possession of the notation of matrix algebra. Problems in counter-examples, complex numbers, irrational numbers, vector algebra, simultaneous equations, co-ordinate geometry, and trigonometry are tackled by the students, and solved, using matrix notation—in many cases where we, their teachers, had not thought to use matrices at all!

XVI. The Inadequacies of Meta-Languages

There is another reason for using "discovery": in point of fact, you usually *cannot* "tell" the student what to do. You and he do not share a sufficiently precise meta-language.

The distinction between language and meta-language, emphasized by logicians since the turn of the century, is this:* we may use a language in communicating with one another in at least three ways:

---

*Actually, I am here making a somewhat metaphorical use of the word "meta-language," but I believe that the metaphor is revealing.

(*i*) We can use "everyday" language, fraught with ambiguities, and hope for the best (incidentally, we *must* start here, for there is nowhere else to start!).

(*ii*) We can define a *more precise* "official" language using "everyday" language just as the vague English language can be used to build the far more precise rules of chess.

(*iii*) We can now seek to discuss our "official" language. To do this, we cannot *use* the "official" language—*it* is the *subject* of our discussion, not the *medium* whereby we conduct the discussion. This "medium" is the "meta-language."

Now, meta-language is always troublesome, and never as simple and precise as the "official" language. As a result, when older texts tried to *tell* the students how, for example, to "add numbers of unlike sign," they usually gave the following rule: "Subtract the smaller from the larger, and use the sign of the larger."

This rule is not merely vague, it is wrong! Try it on

$$+10 + -3$$

Now, -3 is surely smaller than +10 (since it lies to the left on the number line), and so we must subtract -3 from +10,

$$+10 - -3$$

The result is, of course, +13. This already has the "sign of the larger," so further adjustment is unnecessary. It would also be futile; the work is hopelessly in error. We made the mistake of doing what the rule told us to do.

How do students get correct answers? They follow their own correct intuitive ideas, while *claiming* to be following an incorrect rule which would *not* yield the right answer! This is both confusion and hypocrisy.

A correct rule would involve such complications as: "determine which number has the smaller absolute value; subtract this absolute value from the absolute value of the number having the larger absolute value; use *this* result for your answer if the larger absolute value was that of the positive number, or else use the additive inverse of this result in the contrary case."

Unless we wish to write methematics in legal-document style, or worse, we might better leave it to the students to find the procedure for themselves. That is a very major reason for teaching by discovery.

XVII. Two Applications

As a test of our primitive theory of learning, let us apply it to two other areas, and see if it seems useful:

1. The James Pitman-John Downing Initial Teaching Alphabet[24]

While we cannot, by any means, assess all the causes for difficulty in reading,[25] the intuitive "theory of instruction" used by Madison Project teachers would appear to indicate some considerable virtue in the Pitman-Downing *Initial Teaching Alphabet*. This is an alphabet of 44 symbols, roughly one symbol per sound, that is designed specifically to bridge the gap between the spoken language with which the child comes to school, and the confused written language of adulthood. Many of its features parallel various features of Madison Project materials quite closely, such as:

(*i*) One symbol per idea; in Madison Project usage, for example, the *three different meanings* for the traditional symbol "–" are expressed by three different symbols:

$$^-5,\ 5 - 3,\ ^0(-\ 7) = +\ 7$$

(*ii*) Seeking a notation that is as nearly self-evident as possible (as in the Madison Project use of "□").

(iii) Avoiding "teaching," and leaving it largely up to the child to "crack the code"—*but giving him a code which is within his ability to crack!*

(*iv*) Gradual growth, as in the "learning by successive approximations" discussed earlier; the earlier and simpler cognitive structure based upon I.T.A. gradually gives way to the later and more sophisticated, cognitive structure based upon the usual English alphabet.

(*v*) Conformity to the honest and straightforward approach of the child: for example, I.T.A. reads from left to right, as English fails to do. (In Madison Project notation *we have learned, from the children, to* rewrite

$$y = mx + b$$

instead as

$$(\square \times 3) + 5 = \triangle$$

and so on.)

(*vi*) Because the code is left to the child to decipher, and because this task is within his reach, *intrinsic motivation* should be much higher.

(*vii*) Lurking between absolutely intrinsic motivation and external reinforcement, there is the "almost intrinsic" motivation derived from the fact that since the alphabet is easier, the child progresses more quickly to the point where *reading really is a useful tool.*

2. Science in the Elementary School

Although many diverse approaches to elementary school science are presently under discussion, the great importance of having a reasonable cognitive structure, C_n, from which to build "better" versions. C_{n+1}, C_{n+2} . . ., seems to argue for a very heavy emphasis on the *basic concepts* of physics, chemistry, biology, and geology. If this view is correct, it is quite possible that many current elementary science efforts are putting too little emphasis upon the child's acquisition of these basic concepts, upon which future, more sophisticated cognitive structures can most easily be built. (This has almost the appearance of the "critical period" hypothesis!)

XVIII. Conclusion

We live in a world where knowledge, or, in any event, *facts*, are accumulating at an alarming rate. Moreover, obsolescence of facts, theories, attitudes, and even values is more rapid than one can comprehend. How shall we cope with this? It appears that an educational approach based upon a Piaget sequence of successive cognitive structures.

$$\ldots C_k, C_{k+1}, C_{k+2} \ldots$$

each growing out of the one which preceded it, may be an especially valuable way to regard both "knowledge" and "learning." Those who think that such a view is basically obvious to the point of banality should try to reconcile it with various other theoretical approaches to the study of learning, which in some cases will appear incompatible. There really *is* some valuable content here; the theory is not vacuous.

References and Notes

1. Cf. Bruner, Jerome S., "Needed: A Theory of Instruction," *Educ. Leadership,* **20,** 523-532 (1963).

2. Cf. Davis, Robert B., "The Evolution of School Mathematics," *J. Res. Sci. Teaching,* **1,** 260-264 (1963); Davis, Robert B., "The Madison Project: A Brief Introduction to Materials and Activities," The Madison Project, 1962; Davis, Robert B., "Report on the Madison Project," *Sci. Educ. News* (a publication of the American Association for the Advancement of Science), 15-16 (1962); Davis, Robert B., "Report on the Syracuse University-Webster College Madison Project," *Amer. Math. Mo.,* **71,** 306-308 (1964); Davis, Robert B., *Experimental Course Report/Grade Nine,* Report #1, June, 1964 (available from The Madison Project).

3. Cf. Davis, Robert B., "Report on Madison Project Activities, September 1962-November 1963," A report submitted to the National Science Foundation, December 1963.

4. Cf. the films *Experience with Estimating and Measuring Angles* and *Experience with Angles and Rotations.*

5. An earlier, and somewhat different, version of this lesson, with a 6th grade class, can be viewed on the film entitled *Weights and Springs.* A generally similar kind of lesson can be viewed on the film *Average and Variance.*

6. However, the class will have had plenty of prior experience with *functions* and *graphs.* Cf. the films *Experience with Linear Graphs, Second Lesson, Postman Stories, Circles and Parabolas,* and *Guessing Functions.*

7. Cf. the film *Matrices,* and the accompanying pamphlet, which is also entitled *Matrices.*

8. Cf. Davis, Robert B., *Discovery in Mathematics: A Text for Teachers,* Addison-Wesley, Reading, Massachusetts, 1964, pp. 8-15.

9. Cf. the film *Sequencing and Elementary Ideas.*

10. Cf. Bruner, Jerome S., J. J. Goodnow, and G. A. Austin, *A Study of Thinking,* Wiley, New York, 1956, pp. 82 ff. These same pages are also suggestive in relation to our notion of "degree of autonomous control," as discussed on page 152 of the present report.

11. Cf., for example, Whitehead, Alfred N., *Aims of Education* (Mentor Books), Macmillan, New York, 1929.

12. See, for example, the film *Creative Learning Experiences.*

13. Cf., however, Eves, Howard, and Carroll V. Newsom, *The Foundations and Fundamental Concepts of Mathematics,* Holt, Rinehart, & Winston, 1964; Newman, J. R., and E. Nagel, *Gödel's Proof,* New York University Press, New York, 1960.

14. Cf. Sanders, W. J., "The Use of Models in Mathematics Instruction," *The*

Arithmetic Teacher, **11,** 157-165 (1964). For the specific use of "postman stories," cf. the film entitled *Postman Stories.*

15. Cf. Graham, W. A., "Individualized Teaching of Fifth and Sixth-Grade Arithmetic," *The Arithmetic Teacher,* **11,** 233-234 (1964); Mearns, Hughes, *Creative Power,* Dover, New York, 1958; Neill, A. S., *Summerhill,* Hart, Hart, New York, 1960; Brecher, Ruth, and Edward Brecher, "Gifted Children Need Freedom to Learn," *Parents' Magazine,* **37,** 44 ff. (1962).

16. Cf. Bruner, Jerome S., *On Knowing: Essays for the Left Hand,* Harvard University Press, Cambridge, Massachusetts, 1963, pp. 12-15, 18-20. Professor Bruner's remarks also call to mind Leonard Bernstein's delightful essay on the greatness of Beethoven ("Why Beethoven?," an essay included in the volume *Seven Arts, No. 2,* ed. Puma, Permabooks (Doubleday), Garden City, New York, 1954, pp. 33-40).

17. Cf. Boulle, Pierre, *The Test,* Popular Library, New York, 1960.

18. Cf. also Henry, Jules, "American Schoolrooms: Learning the Nightmare," *Columbia University Forum,* **6,** 24-30 (1963).

19. Cf. Davis, Robert B., "Math Takes a New Path," *The PTA Magazine,* **57,** 8-11 (1963), and the film entitled *Education Report: The New Math.*

20. Cf. Ausubel, D. P., "Some Psychological and Educational Limitations of Learning by Discovery," *The Arithmetic Teacher,* **11,** 290-302 (1964); cf. also the very fine discussion by Kersh, Bert Y., "Learning by Discovery: What is Learned?" *The Arithmetic Teacher,* **11,** 226-232 (1964).

21. Cf. Davis, Robert B., *Experimental Course Report/Grade Nine,* Report #1, June, 1964 (available from The Madison Project). See especially Appendix E.

22. Cf. Tolman, E. C., *Behavior and Psychological Man,* University of California Press, Berkeley, California, 1958, Ch. 19. See also Sanders, W. J., "The Use of Models in Mathematics Instruction," *The Arithmetic Teacher,* **11,** 157-165 (1964).

23. Cf. the science fiction novel *The Black Cloud,* by the eminent astronomer Fred Hoyle, Harper, New York, 1958.

24. Cf. Gunther, M., "Cracking the Grown-ups' Code," *Saturday Evening Post,* 34-35 (June 20, 1964).

25. Ross, R., "A Description of Twenty Arithmetic Under-achievers," *The Arithmetic Teacher,* **11,** 235-241 (1964).

26. Ardrey, Robert, *African Genesis,* Dell, New York, 1963.

27. Bruner, Jerome S., *The Process of Education,* Harvard University Press, Cambridge, Massachusetts, 1963.

28. Flavell, John H., *The Developmental Psychology of Jean Piaget,* Van Nostrand, Princeton, New Jersey, 1963.

29. *Goals for School Mathematics,* The Report of the Cambridge Conference on School Mathematics, Houghton Mifflin, Boston, Massachusetts, 1963.

30. *Innovation and Experiment in Education,* A Progress Report of the Panel on Educational Research and Development of the President of the United States, March 1964.

31. Torrance, E. Paul, "Creativity," Department of Classroom Teachers, American Educational Research Association of the National Education Association, Research Pamphlet Series, "What Research Says to the Teacher," #28, April 1963.

Exercises: The Madison Project's Approach to a Theory of Instruction

1. What are the differences between a theory of learning and a theory of instruction? How could you tell when you were reading one and not the other?

2. Observers of Davis' lessons often complain that no *teaching* actually took place. How do you define "teaching"? Is it anything one does which causes learning in others? Give a variety of examples of "teaching" illustrating the limits of your definition. What implications do you see for your own teaching?

3. Davis suggests that teachers should remove themselves from the role of the middleman. Explain what he means by this. Give examples of how this might be done for specific topics in mathematics.

4. Davis believes that " 'each child working alone' is *not* an entirely satisfactory situation." Would Piaget agree? Why or why not?

5. If a teacher believes that a child should learn by successive approximations, how can he lead the child to want a better approximation than what he now has? How does Davis' notion of freedom complicate this situation?

6. Compare Davis' position with that of Bruner and that of Gagné. On what points would he agree? Disagree?

7. Is Davis teaching science or mathematics when he works with weights and springs? What is the difference between science and mathematics? Illustrate.

8. Davis recommends that teachers sometimes make inappropriate suggestions as students work on a problem. How can this be justified? Explain. Cite other articles.

9. Talk to teachers. Can you find evidence to support Davis' comments about how children change in 5th, 6th, 7th, and 8th grades? What implications do *you* see for classroom instruction?

10. Construct your own list of fundamental ideas of mathematics. Compare your list with your colleagues' lists. What readings can you cite that would support the teaching of these fundamental ideas at all grade levels? Does Davis support this notion? Explain.

11. What possible justification can Davis give for not saying "good" when a student has successfully solved a problem?

Explain. Do you agree? Why? What readings support Davis' position? What readings refute his position? Explain.

12. Reflect a little on the quote of Zacharias, "Science is a game played against nature; it is not a game played against the teacher." If you accept the advice given, how will you behave in the classroom? Give an example.

13. What do you think would be Davis' reaction to the mnemonic

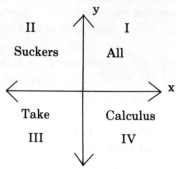

device used to remember which trigonometric functions are positive in each quadrant? Explain. Do you agree? Do Gagné and Davis disagree on this point?

14. Davis lists six "mathematical" objectives and seven more general objectives. Rank these objectives as to their importance to you and justify your ranking. Be sure to give an example of each objective. Is anything missing? Evaluate these objectives.

15. As Davis sees things, what are the primary legitimate motives for students to study mathematics? Do you agree? How should one behave in the classroom if one shares Davis' point of view? Give examples.

Annotated Bibliography

Unit 1: An Introduction to Theories of Mathematical Learning

Bourne, L. E., Jr., Ekstrand, B. R., and Dominowski, R. L. "Contemporary Theories of Concept Formation." In *The Psychology of Thinking*. Englewood Cliffs, N. J.: Prentice-Hall, 1971, 194-222.

Discusses major psychological theories of concept formation.

Bourne, L. E., Jr., Ekstrand, B. R., and Dominowski, R. L. "History and General Theoretical Systems." In *The Psychology of Thinking*. Englewood Cliffs, N. J.: Prentice-Hall, 1971, 20-40.

Introduction to psychological theories of thinking.

Bruner, Jerome. "On Going Beyond the Information Given." In *Contemporary Approaches to Cognition*. Cambridge, Mass.: Harvard University Press, 1957, 41-69.

Discusses learning in terms of coding systems.

Fehr, Howard F., ed. *The Learning of Mathematics, Its Theory and Practice*. 21st Yearbook of National Council of Teachers of Mathematics. Washington, D. C.: National Council of Teachers of Mathematics, 1953.

This book contains specialized articles concerning motivation, sensory learning, language, drill and practice, and problem solving.

Lankford, Francis G., Jr. "Implications of the Psychology of Learning for the Teaching of Mathematics." In *The Growth of Mathematical Ideas, Grades K-12*. 24th Yearbook of National Council of Teachers of Mathematics. Washington, D. C.: National Council of Teachers of Mathematics, 1959, 405-430.

A general view of the psychology of learning applied to mathematics.

McConnell, T. R. "Recent Trends in Learning Theory." In *Arithmetic in General Education*. 16th Yearbook of National Council of Teachers of Mathematics. Washington, D. C.: National Council of Teachers of Mathematics, 1941, 268-313.

A classic article showing application of psychology to teaching arithmetic concepts.

Shulman, Lee S. "Psychology and Mathematics Education." In *69th Yearbook of the National Society for the Study of Education*, 1970, 23-71.

An expanded version of Shulman's psychological controversies in the teaching of science and mathematics.

Unit 2: The Nature of Mathematics Learners

Bearley, Molly and Hitchfield, Elizabeth. *A Teacher's Guide to Reading Piaget.* London: Routledge and Kegan Paul, 1966.

Reprints many of Piaget's reported interviews with children and organizes them in a sequence that is easier to follow than the original reports.

Berlyne, D. E. "Recent Developments in Piaget's Work." *British Journal of Educational Psychology* 27 (1957): 1-12.

Relates Piaget's work to other learning psychologies currently being studied.

Bourne, L. E., Jr., Ekstrand, B. R., and Dominowski, R. L. "Individual Differences Among Problem Solvers." In *The Psychology of Thinking.* Englewood Cliffs, N. J.: Prentice-Hall, 1971, 98-104, 223-234.

Individual differences in learners for problem solving and concept formation.

Bruner, Jerome S., Goodnow, Jacqueline J., and Austin, George A. "Selection Strategies in Concept Attainment." In *A Study of Thinking.* New York: John Wiley & Sons, 1956, 81-96.

A classic study in individual differences in concept learning.

Elkind, David. "The Development of Quantitative Thinking: A Systematic Replication of Piaget's Studies." *Journal of Genetic Psychology* 98 (1961): 37-46.

Discusses experiments designed to investigate the generality of Piaget's work with different children in different cultures.

Elkind, David. "Giant in the Nursery—Jean Piaget." *The New York Times Magazine,* May 26, 1968.

An entertaining, informal introduction to Piaget, the man.

Etuk, Elizabeth. "The Development of Number Concepts: An Examination of Piaget's Theory with Yoruba-Speaking Nigerian Children. Dissertation Abstracts 28: 1295A, No. 4, 1967.

Studies the development of mathematical concepts in children in a radically different culture.

Guilford, J. P. *The Nature of Human Intelligence.* New York: McGraw-Hill Inc., 1967.

Factors of intelligence which have import in diagnosis and planning are identified.

Harrison, D. B. "Piagetian Studies and Mathematics Learning." In *Studies in Mathematics, Vol. XIX: Reviews of Recent Research in Mathematics Education.* Stanford, Calif.: School Mathematics Study Group, 1969.

A good overview and summary of Piaget's theories, with examples of their applications to mathematics curriculum.

Lovell, Kenneth. *The Growth of Understanding In Mathematics: Kindergarten Through Grade Three.* New York: Holt, Rinehart and Winston, Inc., 1971.
Relates formation of elementary concepts to psychological development of the young child.

Lovell, Kenneth E. "Intellectual Growth and Understanding Mathematics." *Journal for Research in Mathematics Education* 3 (May, 1972): 164-182.
Reviews recent research related to Piagetian conservation and the learning of mathematics.

Piaget, Jean. *Genetic Epistemology.* New York: Columbia University Press, 1970.
Explains the logical and psychological foundations underlying Piaget's work. Defines "genetic epistemology" and closely relates it to basic mathematical structures.

Piaget, Jean. *Science of Education and the Psychology of the Child.* New York: The Viking Press, 1970.
Piaget's own interpretation of his work as it applies to education.

Rosskopf, Myron F., Steffe, Leslie P., and Taback, Stanley. *Piagetian Cognitive-Development Research and Mathematical Education.* Washington, D. C.: National Council of Teachers of Mathematics, 1971.
A collection of essays relating mathematics and Piagetian psychology. Particular attention is given to the development of ability in logical thinking.

Siegel, Irving E., *et al. Logical Thinking in Children.* New York: Holt, Rinehart and Winston, 1968.
The essays in this collection report experiments concerning Piagetian developmental psychology. Many of the experiments concern numerical and geometrical concepts.

Unit 3: The Nature of Learning Processes in Mathematics

Bart, William M. "Mathematics Education: The View of Zoltan Dienes." *School Review* 78 (May, 1970): 355-372.
This article relates Dienes' analysis of learning processes to commonly held philosophical positions relative to the foundations of mathematics.

Brownell, William A. and Moser, Harold E. "Meaningful vs. Mechanical Learning: A Study in Grade III Subtraction." *Duke University Studies in Education* 8 (1949): 1-207.
A classic experiment in meaningful instruction.

Gagné, R. M. "The Acquisition of Knowledge." *Psychological Review* 69 (1962): 355-365.

Detailed descriptions of Gagné's experiments with knowledge hierarchies.

Hadamard, Jacques. *The Psychology of Invention in the Mathematical Field.* New York: Dover Publications, 1945.
An analysis of creative processes in the field of mathematics.

Hillman, Bill W. "The Effect of Knowledge of Results and Token Reinforcement on the Arithmetic Achievement of Elementary School Children. *The Arithmetic Teacher* 17 (1970): 676-682.
Suggests that feedback and encouragement are important factors in learning.

Kilpatrick, Jeremy. "Problem Solving and Creative Behavior in Mathematics." *Studies in Mathematics, Vol. XIX: Reviews of Recent Research in Mathematics Education.* Stanford, Calif.: School Mathematics Study Group, 1969, 153-187.
Reviews research related to the complex learning patterns involved in problem-solving.

Luchins, Abraham S. and Luchins, Edith H. *Wertheimer's Seminars Revisited Problem Solving and Thinking, Volumes I, II, III.* Albany, New York: Faculty-Student Association, State University of New York at Albany, 1970.
The penetrating analysis of mind set provides a contrasting view of Ausubel's concept of advance organizer.

Mayer, Richard E. *Dimensions of Learning to Solve Problems.* Ann Arbor, Michigan: Human Performance Center, Department of Psychology, University of Michigan, October, 1972.
Several factors affecting problem solving and transfer of problem solving skills are identified.

Polya, George. *How to Solve It.* Garden City, New York: Doubleday Anchor Books, 1957.
Presents the role of heuristics in learning to solve mathematics problems.

Suppes, Patrick. "Mathematical Concept Formation in Children." *American Psychologist* 21 (1966): 139-150.
Considers learning as an "all-or-nothing" phenomenon, rather than a gradual process.

Wertheimer, Max. *Productive Thinking.* (Enlarged edition). New York: Harper & Row, Publishers, 1959.
The classic rendering of Gestalt psychology in the field of mathematics education.

Unit 4: Instructional Applications

Ausubel, David P. *Educational Psychology: A Cognitive View.* New York: Holt, Rinehart and Winston, 1968.

This book provides a more extensive analysis of discovery learning than the article included in this book of readings.

Bruner, Jerome. Some theorems on instruction illustrated with reference to mathematics. *Theories of Learning and Instruction, 63rd Yearbook, Part I, National Society for the Study of Education.* Chicago: The University of Chicago Press, 1964, 306-335.
Discusses the implications of cognitive psychology for classroom practice.

Dienes, Z. P. and Golding, E. W. *Exploration of Space and Practical Measurement: Learning Logic, Logical Games: Set, Numbers and Powers.* New York: Herder and Herder, 1966.
Illustrates activities and games used with children for teaching significant mathematics at an early age.

Engelmann, Siegfried. *Conceptual Learning.* Palo Alto, Calif.: Fearon Publishing Co., 1969.
Describes in detail the task of building a hierarchial analysis of teaching.

Higgins, Jon L. *Mathematics Teaching and Learning.* Worthington, Ohio: Charles A. Jones Publishing Co., 1973.
Examines major psychological theories and investigates their applications to classroom teaching.

Kidd, Kenneth P. "Improving the Learning of Mathematics." *The Mathematics Teacher* (October, 1954): 393-400.
Suggests several teaching strategies and techniques based on general psychological principles.

Mager, R. F. *Developing Attitude Toward Learning.* Palo Alto, Calif.: Fearon Publishing Co., 1968.
A behavioral objective approach to attitudes (affective domain).

Mager, R. F. *Preparing Objectives for Programmed Instruction.* Palo Alto, Calif.: Fearon Publishing Co., 1968.
A short introduction to the writing of behavioral objectives. Judged superficial by some; a classic by others.

Riedesel, C. Alan. "Verbal Problem Solving: Suggestions for Improving Instruction." *The Arithmetic Teacher, II* (1964): 312-316.
Discusses an instructional program designed to teach problem solving in mathematics.

Romberg, T. A. and Wilson, J. W. "The Development of Mathematics Achievement Tests for the National Longitudinal Study of Mathematical Abilities." *The Mathematics Teacher* (May, 1968): 489-495.
Illustrates a taxonomic approach to the development of objectives (test items).

Skemp, Richard R. *The Psychology of Learning Mathematics.* Middlesex, England: Penguin Books Ltd., 1971.

Develops the idea of schema as an underlying mechanism in the learning of mathematics.

Thorndike, E. L. *The Psychology of Algebra.* New York: Macmillan Co., 1923.

One of the first attempts to apply psychology to the teaching of mathematics.

Van Engen, Henry. "Some Psychological Principles Underlying Mathematics Instruction." *School Science and Mathematics* (April, 1961): 242-250.

Relates curriculum selection and construction to basic psychological principles.